2ND MARINE AIR WING

Transitioning "The Fight Tonight Force"

ROBBIN LAIRD

Copyright © 2025 by Robbin Laird All rights reserved.

No portion of this book may be reproduced in any form without written permission from the publisher or author, except as permitted by U.S. copyright law.

Library of Congress Control Number: 2025909807

The cover photo was taken by the author when onboard the HMS Illustrious in 2007.

This book is dedicated to LtGen "Dog" Davis with whom I have had many conversations about Marine Corps air over the years, when he was in the Pentagon, when he was in North Carolina, when he was onboard the USS Wasp and in Canberra, Australia when he was giving a presentation at the Sir Richard Williams Foundation.
His intellect, his service and his leadership have been and remain outstanding for the USMC and for the nation.

From left to right, U.S. Marine Corps LtCol Jeffrey C. Davis. left, the outgoing commanding officer of Marine Fighter Attack Squadron (VMFA) 314, Marine Aircraft Group 11, 3rd Marine Aircraft Wing, retired LtGen Jon M. Davis, center, and Maj Eric Davis, right, an air officer with Headquarters and Service Battalion, Marine Corps Forces Pacific, pose for a photo during a change of command ceremony at Marine Corps Air Station Miramar, California, March 21, 2025.

Contents

Introduction	vii
1. 2nd MAW Today	1
2. First Encounter: The USMC on HMS Illustrious, 2007	9
3. The Evolution of 2nd Marine Aircraft Wing: Leadership Perspectives Over the Past Two Decades	13
4. 2009 Visit: Marine Air in the Land Wars	43
5. 2010 Visit to New River: An Osprey Update	58
6. 2011 Visit to New River: USMC Con-Ops in Evolution	62
7. 2012 Visits: Focusing on the Return to the Sea	73
8. 2013 Visit: Ospreys, Harriers and the Future	86
9. 2014 Visit: Reshaping USMC Expeditionary Capabilities	109
10. 2015 Visit: Strands of Transition	139
11. 2019 Visits: Transitioning to Arctic Operations	160
12. 2020 Visit: Working the Integrated Distributed Insertion Force	182
13. 2021 Visit: New Operational Concepts	200
14. 2024 Visit: F-35s and Distributed Operations	218
15. 2025 Visit: Working Distributed Air Operations	230
About the Author	243
About the Contributors	247
About Our Websites	251
USMC Transformation Series: A Comprehensive Analysis	255
Notes	259

Introduction

From the Marine Corps base at Cherry Point to the skies above the Arctic Circle, the 2nd Marine Aircraft Wing stands as one of America's most adaptable and forward-leaning military aviation units.

For over a decade, I've had the privilege of observing this remarkable organization as it has transformed itself from a force engaged in Iraq and Afghanistan to one preparing for the complex challenges of distributed maritime operations in contested environments.

This book chronicles my fifteen-year journey alongside the Marines of 2nd MAW, capturing their evolution through firsthand accounts, interviews with commanding officers, and observations of their training and operations.

My relationship with 2nd MAW began in 2007 when I flew aboard the HMS Illustrious via an Osprey, departing from the Pentagon helipad to observe Marines working with the Royal Navy, operating USMC Harriers off a jump-jet carrier as part of a training exercise. This unique event marked the first time a non-U.S. ship had received an Osprey landing.

What began as a fascinating experience aboard the HMS Illustrious would evolve into a years-long journey of interaction and observation, following the Marines through their airpower transformation from that moment to the present day.

Through the chapters that follow, you'll witness the remarkable evolution of Marine aviation capabilities, from the revolutionary MV-22 Osprey that redefined operational reach to the fifth-generation F-35 Lightning II that transformed battlefield awareness.

You'll hear directly from the commanding generals who led 2nd MAW from 2011 to 2025, each bringing unique perspectives on readiness, innovation, and the core Marine mission of supporting infantry on the ground.

Throughout the fifteen-year journey documented in these pages, one theme remains constant across all commanding generals: the unwavering commitment to supporting Marines on the ground. As Major General Davis put it in 2012: *The Marine ACE exists for one reason, to make our Marines better fighters. When we improve the ACE's capability to fight, we increase the warfighting effectiveness of our riflemen.*

From operations in Afghanistan's challenging terrain to exercises in Norway's Arctic conditions, from the integration of revolutionary aircraft to the development of distributed operations concepts, this book tells the story of an organization that has continuously adapted to changing threats while maintaining its fundamental purpose: making Marines on the ground more effective and keeping them alive in combat.

The 2nd Marine Aircraft Wing truly embodies what it means to be "The Fight Tonight Force," one that is perpetually ready, consistently innovative, and utterly committed to their fellow Marines. This is their story.

I would add that the photo below taken during a 2024 visit, highlights how much pleasure it really is to talk to both young and older more experienced Marines when I come to one of the three bases operated by 2nd MAW.

These Marines in the photo below had recently come back from an exercise in the Nordic region, a region which I have often visited and not only did I learn a great from the Marines but I could share some stories from my visits to the Nordics as they were reshaping their defense capabilities.

Credit: 2nd MAW

The photos in the book are credited to the USMC as indicated. If there is no credit given then they were taken by me.

The articles were mostly written by me, and hence there is no name following the title of an article in those cases. But two persons contributed to my visits from time to time, and in those cases the co-author is clearly indicated.

Ed Timperlake whose life-long experience of the Marines both as a squadron leader and pilot and his knowledge of the challenges faced by Marines has been a key partner in my own journey to understand the evolution of the modern USMC.

In addition, Murielle Delaporte, a noted military analyst, accompanied me to some of my interviews and her help was indispensable in those visits.

ONE

2nd MAW Today

On July 1, 2025, I had a chance to talk with the Public Affairs Officer of 2nd Marine Air Wing. Major Joseph Leitner. Major Leitner provided me with an overview on the structure of the Wing and its locations as of July 1, 2025.

The 2nd Marine Air Wing (2nd MAW) represents one of the most geographically distributed yet operationally integrated aviation commands in the U.S. military.

Spanning two states along the Eastern Seaboard, this 11,500-strong force of Marines and sailors operates from three strategically positioned air stations, each serving distinct but complementary roles in America's crisis response capabilities.

Based on my discussion with Major Leitner, here are the 2nd MAW units clustered by their geographic locations within the 2nd Marine Air Wing:

Marine Corps Air Station Cherry Point, North Carolina

- Marine Wing Headquarters Squadron-2 (MWHS 2)
- Marine Aircraft Group 14 (MAG-14)

- VMFA-542 (F-35B Lightning II)
- VMFA-251 (F-35C Lightning II)
- VMA-223 (AV-8B Harrier II - last operational Harrier squadron, standing down in 2026)
- VMA-231 (Administrative status only, concluded operations May 2025, officially standing down September 2025)
- VMGR-252 (KC-130J Super Hercules)
- VMU-2 (MQ-9 Reaper fleet replacement squadron)
- KC-130J Fleet Replacement Detachment (FRD)
- MALS-14 (Marine Aviation Logistics Squadron)
- Marine Air Control Group 28 (MACG-28)
- Marine Wing Communications Squadron 28
- MACS-2 (Marine Air Control Squadron 2)
- 2nd LAAD (Low Altitude Air Defense Battalion)
- MASS-1 (Marine Air Support Squadron 1 - controllers and air traffic control)
- MWSS-271 (Marine Wing Support Squadron 271)

With regard to Marine Wing Headquarters Squadron-2 (MWHS 2), it serves as the headquarters formation for 2nd MAW. All of the staff sections that report to the General and support the Wing's general operations are contained in this formation.

These are the less glamorous roles, but they are nonetheless vital to the Wing's functioning. For example, a large portion MWHS-2 deployed to Norway to serve as the MAGTF Command Element during Nordic Response 24.

In addition, there are the major staff sections contained in the wing headquarters:

- Operations (G-3)
- Administration/Manpower (G-1)
- Intelligence (G-2)
- Logistics (G-4)
- Communications (G-6)
- Finance (G-8)

- Aviation Logistics (ALD)
- Department Safety and Standardization
- Inspector General
- Wing Chaplain
- Wing Surgeon
- Wing Band
- Communication Strategy and Operations (Public Affairs)
- Staff Judge Advocate (Legal Advisor)

Marine Corps Air Station New River, North Carolina
Marine Aircraft Group 26 (MAG-26) - All MV-22B Ospreys

- VMM-162
- VMM-266
- VMM-365
- VMM-261
- VMM-263
- VMMT-204 (Fleet replacement/training squadron)
- VMM-264 (Set to reactivate early 2026)
- MALS-26
- Marine Aircraft Group 29 (MAG-29) - CH-53 and H-1 aircraft
- HMH-461 (CH-53K King Stallion - operational squadron)
- HMH-464 (CH-53E Super Stallion)
- HMHT-302 (CH-53 fleet replacement/training squadron)
- HMLA-167 (AH-1Z Viper and UH-1Y Venom)
- HMLA-269 (AH-1Z Viper and UH-1Y Venom)
- Other New River Units: MWSS-272 (higher headquarters is MACG-28)

Marine Corps Air Station Beaufort, South Carolina

- Marine Aircraft Group 31 (MAG-31)

- VMFA-312 (F/A-18 Hornet - transitioning to F-35B in FY28, moving to Cherry Point)
- VMFA-224 "Bengals" (F/A-18 Hornet - receiving F-35B in FY26)
- VMFA-533 (F-35C Lightning II - first operational F-35 squadron)
- VMFAT-501 (F-35 East Coast fleet replacement squadron)
- VMFA-115 (Currently stood down, will reactivate as F-35C squadron at Cherry Point)
- MALS-31
- MWSS-273 (higher headquarters is MACG-28).

Marine Corps Air Station Cherry Point in North Carolina serves as the nerve center of 2nd MAW operations, housing both the wing's most advanced strike capabilities and its critical support infrastructure. The base hosts two major groups that exemplify the modern Marine Corps' evolution toward distributed operations and technological superiority.

Marine Aircraft Group 14 (MAG-14) at Cherry Point represents the cutting edge of Marine aviation technology. The group operates the service's newest platforms, including F-35B and F-35C Lightning II squadrons VMFA-542 and VMFA-251. These fifth-generation fighters provide the wing with advanced stealth and sensor capabilities essential for operations in contested environments.

I have been engaged with the introduction of the F-35 beginning in 2004 when working for Secretary Mike Wynne, first when he was head of defense acquisition and then as Secretary of the Air Force. I just published a new edition of my book entitled, *My Fifth Generation Journey* which tells this narrative.

Perhaps more significantly for the wing's global reach, Cherry Point houses VMGR-252 flying KC-130J Super Hercules aircraft. As Major Joseph Leitner explained: *That's our organic lift capability. If the wing needs to go anywhere, especially on rapid notice, we have 252 organically.*

This capability proved crucial during recent rapid deployments, including moving elements to Guantanamo Bay following presidential orders and supporting operations in Panama.

The base also serves as home to Marine Air Control Group 28 (MACG-28), which Leitner describes as a "Swiss Army knife" in supporting the wing. This diverse group encompasses everything from communications and air traffic control to low altitude air defense and the Marine Wing Support Squadrons (MWSS) that provide expeditionary airfield construction and forward arming and refueling points (FARPs).

Located adjacent to Camp Lejeune, Marine Corps Air Station New River houses the preponderance of 2nd MAW's assault support aircraft. New River's two aircraft groups reflect the Marine Corps' commitment to vertical lift capabilities.

Marine Aircraft Group 26 (MAG-26) operates exclusively MV-22B Ospreys across seven squadrons, including the fleet replacement squadron VMMT-204.

The Osprey's unique tiltrotor design provides the Marine Corps with unprecedented speed and range for assault support missions, bridging the gap between traditional helicopters and fixed-wing transport. I have published two books in 2025 which tell the tiltrotor story since its first deployment.

Marine Aircraft Group 29 (MAG-29) combines the Marine Corps' heavy lift and light attack helicopter capabilities. The group operates both CH-53E Super Stallions and the new CH-53K King Stallions, with HMH-461 serving as the first operational King Stallion squadron. The two light attack helicopter squadrons, HMLA-167 and HMLA-269, provide close air support with their AH-1Z Vipers and utility support with UH-1Y Venoms.

I have published a book as well on the coming of the CH-53K which highlights the nature of the innovative digital aircraft and its impact on the future force.

The CH-53K working with the KC-130J will form a key support capability for the Marines as the hone their capabilities for distributed operations. The KC-130J squadron at Cherry Point is located nearby the new CH-53K squadron at New River, which enables them to experience working with the new generation heavy lift helicopter.

Both the CH-53K and KC-130J use the 463L-standard air cargo pallets, which offers immense advantages when it comes to loading and unloading materiel. This shared pallet system enables tail-to-tail loading operations. The tail-to-tail transfer of supplies allows distribution of sustainment in the minimum time period of vulnerability by reducing break-bulk requirements

Marine Corps Air Station Beaufort in South Carolina currently serves as a critical transition point in the Marine Corps' aviation modernization. Marine Aircraft Group 31 (MAG-31) operates F/A-18 Hornets while simultaneously managing the complex transition to F-35 platforms.

I followed the F-35Bs which were stood up at Eglin AFB and then came to Beaufort in 2015. This was when key sea trials on the USS Wasp were conducted paving the way for declaration of IOC for the F-35B.

VMFA-533 stands as the first operational F-35B squadron aboard MCAS Beaufort, while VMFAT-501 serves as the East Coast fleet replacement squadron for F-35 training. The base represents both the present and future of Marine tactical aviation, with squadrons like VMFA-312 and VMFA-224 scheduled to transition from Hornets to Lightning IIs over the coming years.

VMFA-251 at Cherry Point is 2nd MAW's first F-35C squadron. VMFA-251 received its first F-35C's in September 2024.

Cherry Point's position in central North Carolina places it within easy reach of both Norfolk Naval Base and the Research Triangle, providing access to naval aviation support and defense industry partnerships.

New River's co-location with Camp Lejeune ensures seamless integration between aviation and ground elements.

Beaufort's South Carolina location extends the wing's geographic footprint while providing training airspace over the Atlantic, and in places like the Townsend Bombing Range in Georgia.

This geographic structure directly supports 2nd MAW's designation as the Marine Corps' Crisis Response Force.

Unlike other Marine Expeditionary Forces aligned to specific combatant commands, 2nd MAW maintains the flexibility to respond globally. Recent operations spanning from SOUTHCOM to EUCOM to AFRICOM demonstrate how the three-base structure enables rapid response across multiple theaters. The wing's organic lift capability, centralized at Cherry Point, can quickly move personnel and equipment from any of the three bases to global crisis points.

Meanwhile, the specialized capabilities at each location from New River's assault support aircraft to Beaufort's strike fighters provide commanders with a full spectrum of aviation options.

The ongoing transition to F-35 platforms will reshape this geographic distribution over the coming years. Several squadrons currently at Beaufort are scheduled to move to Cherry Point as they transition to new aircraft, potentially concentrating more strike capability at the northern base while Beaufort focuses on training and transition missions.

The planned reactivation of squadrons like VMA-231 and VMM-264 will add capability without necessarily changing the fundamental geographic structure. Instead, these additions will deepen the bench strength at each location while maintaining the strategic distribution that has proven so effective for crisis response operations.

The 2nd Marine Air Wing's three-base structure represents more than administrative convenience. It embodies a strategic approach to crisis response that balances operational flexibility with geographic resilience. From Cherry Point's command and strike capabilities to New River's assault support concentration to Beaufort's transition management, each base contributes unique capabilities to a unified whole.

This geographic foundation enables 2nd MAW to fulfill its role as the Marine Corps' premier crisis response aviation element, ready to project power and provide support wherever the nation's interest's demand. As the wing continues evolving with new platforms and capabilities, this three-base structure will likely remain the geographic backbone of Marine aviation's eastern seaboard operations.

The story of 2nd MAW was written in contrails and engine noise, in the dedication of Marines and sailors who understood that their mission was not just about flying aircraft but about projecting American power and values to every corner of the globe where American interests needed defending.

TWO

First Encounter: The USMC on HMS Illustrious, 2007

My first encounter with 2nd Marine Air Wing was when I flew onboard the HMS Illustrious via an Osprey. That was in 2007 when we left the Pentagon helo pad and then landed on the HMS Illustrious off the Virginia Coast.

I was working as a consultant in the Pentagon with Michael W. Wynne first in his role as the acquisition chief and then as Secretary of the USAF. I started at that time to engage with Marines and this when I first met with LtGen Trautman, Deputy Commandant of Aviation.

In 2007, the HMS Illustrious was the first non-U.S. ship on which an Osprey was to land.

I had the opportunity to be aboard one of those Ospreys and land on the ship and observe Marines working with the Royal Navy and the Marines were operating their Harriers off of the jump-jet carrier as part of their training effort.

At the time, British Harriers were operating in Iraq and were not aboard the ship itself. The Wikipedia entry about the event reads as follows: *United States Marine Corps (USMC) AV8B Harriers conduct fixed wing work on HMS Illustrious ahead of Operation Bold Step. At the rear of the flight deck, an Osprey MV-22 aircraft can just be seen.*

AV8B Harrier Jets from the United States Marine Corps (USMC) Combined Marine Harrier Force have begun a fixed wing work up period on board HMS Illustrious ahead of the US-led Joint Task Force Exercise (JTFX), Operation Bold Step.

The AV8B Jets are on board ILLUSTRIOUS as part of a coalition exercise taking part between the US Navy, USMC and the Royal Navy. The goal of the exercise is to demonstrate allied interoperability and the expeditionary capabilities of Vertical/Short Take-Off and Landing aircraft. This exercise marks the first time that an American aviation unit of this scale has been embarked aboard a foreign warship and for the British it is the largest ever embarkation of foreign jets on a UK aircraft carrier.

The Jets have been practising ship-borne operations, a skill which is extremely challenging and requires exacting standards from both the pilots and the crew of the 22,500 tonne warship. The principle aim is to qualify in day CVS operations; although air combat, electronic warfare training and day combat ready work-up sorties will also be conducted as part of the USMC's participation for the JTFX.

For the Ship's Air Department it has proved to be an excellent training opportunity, with aircraft handlers, Air Traffic and Fighter Controllers, and the air safety organisation all playing a part in the safe and successful embarkation of the squadron.

The Flight Deck Officer, Lt Jon Llewellyn said "Having the Combined Harrier Force on board for this period has provided excellent training and regeneration opportunities for the Flight Deck teams, integrating with the USMC and building the tempo of the deck safely and efficiently. Having the AV8B jets on board has been a pleasure!"[1]

The experience was fascinating but I had no idea that I was embarking on a journey interacting and observing Marines through their airpower transformation path from then until now.

But I did take a number of photos when onboard the ship which capture this unique moment in aviation history.

First Encounter: The USMC on HMS Illustrious, 2007 • 11

THREE

The Evolution of 2nd Marine Aircraft Wing: Leadership Perspectives Over the Past Two Decades

Throughout the past decade and a half, 2nd MAW has undergone significant transformations in capabilities, operations, and leadership. Yet one theme remains constant across all commanding generals: the unwavering commitment to supporting Marines on the ground.

As Lieutenant General Davis put it in 2012: *The Marine ACE exists for one reason, to make our Marines better fighters. When we improve the ACE's capability to fight, we increase the warfighting effectiveness of our riflemen.*

From the revolutionary impact of the Osprey to the game-changing potential of the F-35B, from operations in Afghanistan to exercises in the Nordic region, 2nd MAW continues to evolve while staying true to its fundamental purpose: making Marines on the ground more effective and keeping them alive in combat.

Modernization and Innovation at the 2nd Marine Aircraft Wing: A Conversation with Major General Swan

May 6, 2025

U.S. Marine Corps MajGen William Swan, the commanding general of 2nd Marine Aircraft Wing, adjusts his uniform before climbing into an F/A-18D Hornet aircraft with Marine Fighter Attack Squadron 312, Marine Aircraft Group 12, 1st MAW during exercise Cope North 25 at Andersen Air Force Base, Guam, Feb. 11, 2025. (U.S. Marine Corps photo by Cpl. Dahkareo Pritchett)

During my latest visit to 2nd Marine Wing, I had a chance to talk with Major General Swan, Commanding General of the 2nd Marine Aircraft Wing (2nd MAW) on 29 April 2025.

He offered insights into how the wing is navigating modernization while maintaining readiness for global operations. With a focus on integrating new aircraft systems, enhancing maintenance capabilities, and fostering a culture of innovation, the 2nd MAW is positioning itself to meet the challenges of modern warfare.

Major General Swan highlighted significant progress in modernizing the wing's aircraft inventory. The F-35 squadrons are "coming faster now thanks to improved delivery timelines," with VMFA-542 "up on step and ready to go."

The squadron participated last year in Exercise Nordic Response 24, during which it operated the first U.S. F-35's in Sweden and rehearsed distributed aviation operations in the high north. Swan also noted the reactivation and first F-35C deliveries to VMFA-251 in late 2024, and the first F-35B deliveries to VMFA-533 in October. Each were important milestones in 2nd MAW's tactical aircraft modernization efforts.

The CH-53K heavy-lift helicopter program is also advancing, bringing transformative capabilities to the Marine Corps. Major General Swan emphasized the helicopter's impressive lift capacity and fly-by-wire technology which enables precise hovering over loads. At a recent Service Level Training Exercise (SLTE), the CH-53K lifted a fully combat loaded Light Armored Vehicle for the first time, demonstrating its capabilities to Marines in the ground combat element and allowing them to experience those capabilities firsthand.

I think the future ACE [Aviation Combat Element], if you will, is going to be more connected, more capable, more lethal, Swan noted. This modernization extends to attack helicopters, which are receiving Link 16 data links to enhance connectivity with F-35s and other platforms.

While advancing modernization efforts, the 2nd MAW must maintain operational readiness for global force management commitments in the Pacific, Europe and Africa. As the Marine Corps' service-retained ACE, it is also tasked to be ready to respond to crisis or contingencies globally, in any geographic combatant command, as opposed to singularly focusing on one theater or another. This creates a complex challenge for leadership.

Swan described 2nd MAW's force generation cycle, noting that while some units are deployed, others are either preparing for deployment or recently returned. He noted that the wing pays close attention to ensuring that readiness to deploy is managed appropriately while also balancing modernization efforts across squadrons that are transitioning to new aircraft.

Swan's priorities for the wing are straightforward: *Be ready. Take care of our people. Find more cowbell.*

He emphasizes that Marines must be trained to execute their assigned missions, whether for global force management or crisis response. However, he acknowledges the challenges posed by program delays, noting that new capabilities are sometimes delayed.

Perhaps most revealing is Major General Swan's approach to innovation, which he calls "more cowbell" which is a reference to the famous Saturday Night Live skit. He distributes actual cowbells to Marines who develop innovative solutions to persistent problems. Since implementing the cowbell award program last summer, Swan has handed out more than forty cowbells to deserving Marines who innovated or improved capabilities at the unit level.

The Marines want to do a great job, and they want to be better. They want to win, Swan explained. This philosophy encourages personnel to constantly "improve your position" and find better ways to accomplish the mission.

One of the wing's most promising innovation areas is predictive maintenance. Swan described efforts to leverage aircraft sensor data and artificial intelligence to predict component failures before they occur. This approach aims to shift from unscheduled to scheduled maintenance or, in other words, fixing parts before they fail during critical missions.

How do we do scheduled maintenance? Meaning, hey, I know that at two more hours, this generator, this servo cylinder, this radio is going to break, and I need to fix it now so I can send it on a 10-hour mission, Swan explained.

This capability would be particularly valuable in contested logistics environments, allowing maintenance to occur *at a time and a place of your choosing, vice an inopportune time where you put people at risk and the mission at risk.*

Swan believes combining government data resources with AI algorithms could revolutionize maintenance and supply chains, creating *a better, more capable force that can iterate and turn and decide inside the OODA loop of the enemy.*

While the broader military faces recruitment and retention challenges, Major General Swan reports that the Marine Corps is "nailing retention." He attributes this success to the Corps' commitment to maintaining high standards.

We haven't lowered our standards, and are proud of that, in fact, unapologetic about maintaining our standards and our people, Swan said.

The 2nd MAW's approach to retention focuses on team building and mastering fundamentals. *My philosophy... we build a team of teams, and we take care of our Marines. We are brilliant at the basics,* Swan explained. This leadership philosophy creates an environment where Marines feel valued and part of a winning organization.

When discussing the Aviation Combat Element's importance to Marine Air-Ground Task Force (MAGTF) operations, Swan underscored the following: *The ACE, the air wing, is the center of gravity for the MAGTF, for the ability to maneuver, the ability to [deliver] long-range fires and provide combined arms effects for the maneuver element.*

This view positions aviation as the essential enabler for ground operations, particularly in distributed operations across contested environments. Swan emphasized that each echelon has its own center of gravity — for aircraft groups, it's the maintenance logistics squadrons; for the wing, it's the command-and-control group.

As the 2nd MAW continues its modernization journey, the integration of digital systems, predictive maintenance, and advanced platforms like the F-35 and CH-53K will reshape Marine aviation capabilities. Major General Swan's leadership approach, combining readiness with innovation, provides a framework for managing this complex transition.

The challenges remain significant ones ranging from maintaining readiness with limited amphibious shipping to accelerating the integration of new technologies. However, the focus on building teams, empowering innovation, and leveraging emerging technologies positions the 2nd MAW to meet these challenges while delivering combat power when and where it's needed.

As Swan succinctly put it: *How do we go faster and get better and more lethal?*

MajGen Benedict on His Time as CG of 2nd Marine Air Wing

May 22, 2024

U.S. Marine Corps MajGen Scott F. Benedict, third from left, commanding general of 2nd Marine Aircraft Wing, speaks with NATO service members during Exercise Nordic Response 24 in Alta, Norway, March 6, 2024. (U.S. Marine Corps photo by Lance Cpl Christian Salazar)

On May 14, 2024, I visited 2nd Marine Air Wing to speak with Commanding General MajGen Scott Benedict, two days before his change of command and retirement after a distinguished Marine Corps career.

I had last visited 2nd MAW when LtGen Michael Cederholm was commander (now I MEF Commander). At that time, I focused on the historic opportunity for Norfolk fleet and North Carolina Marines to work through Nordic integration to create unprecedented defense capacity.

During my 2021 visit, I had written about the transformation potential: The North Carolina-based Marines have equipment prepositioned in Norway and exercise frequently with Norwegians. Through the Cold War and beyond, those Marines had the mission to reinforce Norway in crisis. But in an era of stated Marine Corps-Navy integration, this could change dramatically.

The geography highlights the growing role of the High North. If one examines the map, enhanced Nordic integratability can reshape how Marines might reinforce the air-sea battle where the U.S. Navy is heading in its reset to fight and prevail in the 4th Battle of the Atlantic. Given Navy priority concerns regarding Murmansk and the Kola Peninsula, allies best positioned to reinforce U.S. efforts are crucial. Iceland, Denmark (Faroe Islands and Greenland), Norway, Sweden and Finland anchor effective regional deterrence and can shape any Fourth Battle of the Atlantic outcome.

MajGen Benedict and 2nd MAW spent the last two years building Scandinavian relationships, culminating in the recent Nordic Response Exercise with Sweden and Finland participating as full NATO members. For the USMC, Swedish and Finnish NATO entry means substantial change from primarily bolstering Norway in crisis to working with all Nordic forces which is a capability they demonstrated by exercising with all three countries.

This allows the USMC to embed its Marine Air Ground Task Force integrated capabilities within the Nordic defense network and support the U.S. fleet operating in regional defense. The air capabilities — Ospreys, F-35s, F-18s, and CH-53Ks — can operate from land to support the fleet or from fleet to support land-air operations. This dual capability is truly unique and what the USMC brings to the fight.

The extensive training and integration with all Nordic countries represents a very significant development, and lessons learned in this key North Atlantic defense region can apply to the Pacific as well.

MajGen Benedict provided a wide-ranging picture of Marine Corps activities shaping this concept of operations and 2nd MAW's key role working with 2nd and 6th Fleets regionally.

He started by underscoring his view of USMC-Naval integration: *I went to a senior commanders' course in Naples focusing on maritime combined arms operations. It struck me that both Navy and Marines almost solely focus on Marine capabilities being employed from the sea, but not so much on how we can come from the land to support the naval campaign.*

The opportunity to work with the Nordics as they continue enhancing defense integration clearly allows us to demonstrate and take advantage of that opportunity and shape innovative ways to do so. We did that in Nordic Response 2024 as well. There's a lot we can achieve in littoral operations without solely operating from an amphibious ship.

During the exercise, working with Nordic air chiefs, Benedict emphasized that Marines and American forces are working closely on shaping effective C2 across the coalition force to operate as integrated as possible.

Notably, 2nd MAW brought its first F-35 squadron to the exercise. With Norwegians already operating F-35s, Denmark and Finland following, plus F-35s from the UK's Prince of Wales carrier, the F-35s are very interoperable and capable of higher integration levels. Adding German and Polish F-35s will create substantial additional capability as well.

The Finns particularly excel at distributed air operations (DO) on their soil, and Marines worked closely with them. The progress since I last talked to 2nd MAW pilots working with Finns is significant. When I spoke to pilots in 2018, they indicated Finns were teaching them about DO. Now Marines are working hard on their own DO approach, and having Finland which has lived in a big power's shadow for so long makes them a key partner in evolving F-35 DO.

We discussed how distributed air operations differed from simply being a Forward Arming and Refueling Point (FARP). The difference is profound.

While FARP remains DO's most visible manifestation, it's the final action and doesn't reflect significant choices and work necessary to create that capability. Re-arming air assets at remote, ever-changing locations allows forces to disaggregate for protection then aggregate to mass combat capability rapidly to maintain tempo.

As MajGen Benedict explained: *The difference is the backside. Where is the location? Why that location? How do we get fuel there? How do we get ordinance there? How do we provide force protection? How do we maintain aircraft and for how long? How long are we staying? When do we move? All these things are what I call the backside of distributed operations.*

To address this challenge, 2nd MAW added a functional area inside its Tactical Air Command Center called a Distributed Operations Coordination Cell. According to Benedict: *Here we plan out and initiate coordination and execution of all aviation ground support necessary to support an ATO in a distributed environment.*

Thinking about Marines supporting the Navy from the Nordic landmass raises integration questions and leveraging unique air assets. Beyond the F-35, the Osprey's speed, range, and ability to operate across fleet decks including Military Sealift Command ships is a key enabler when providing Nordic region supplies and support to the fleet.

Benedict's Osprey experience began as SP-MAGTF commander in Spain, where Ospreys worked with KC-130Js throughout African operational space. He commented on subsequent MEU deployment: *Their capability to operate distributed within a theater is amazing. The MEU commander's ability to move force from disaggregated or distributed manner to get the right force at the right time is crucial for the USMC and combatant commander. It is a unique and indispensable capability. Ospreys are the backbone of distributed assault support. No doubt about that. It enables operational maneuver from sea and land to support naval forces.*

The newest addition to 2nd MAW is the CH-53K. Benedict, as a Cobra pilot who didn't often praise other helicopters, emphasized: *I have flown the Kilo and it's an amazing helicopter. With fly-by-wire capability and ability to hover over loads, the stability, reduced aircrew workload — it's a game changer.*

He highlighted the King Stallion's ability to hold hover for extended periods, crucial when it recently lifted a damaged Navy helicopter from a mountain crevice. The Kilo's stability versus the Echo was mission-critical.

The second key capability is internal load-out. Because the CH-53K holds standard USAF pallets internally, it can take cargo from larger transport aircraft like C-17s or C-130s in 'tail-to-tail' fashion, where previously those pallets required breakdown into smaller loads.

To support Marines at remote locations, a pallet with needed materials is cross-loaded to the Kilo at a safe location, then transported long-range (the aircraft is air-refuelable) and landed with rapid pallet removal, minimizing deck time. This significantly reduces operational signature which is a key objective for current USMC operational thinking.

In short, MajGen Benedict's time at 2nd MAW has been historically significant as Marines and Navy work with Nordics to shape a more integrated North Atlantic defense capability.

An Update on 2nd Marine Air Wing from MajGen Cederholm: July 2021

July 29, 2021

The Marines are undergoing a strategic shift with urgency, moving from Middle East land wars to strategic competition readiness. The Marine Corps exists to provide a globally deployable Naval Expeditionary Force, preparing for initial engagements in contested areas and supporting high-end fights within the Joint Force and allied partnerships.

This represents both a strategic shift and shock, transitioning from Middle East combat conditions to diverse mission engagements in Pacific and North Atlantic operations. While II MEF never stopped training to fight in any clime and place including increased Nordic region training with allies, the reality is this generation of Marines has focused on Middle East counter-terror operations, not littoral operations against strategic competitors.

The reset involves mastering hybrid warfare and gray zone operations while enabling Joint Force and allied escalation capabilities as required.

During my July 2021 visit to 2nd MAW, I discussed these challenges with Major General Cederholm. According to Cederholm, *We are approaching readiness levels not seen in decades. On some days, our readiness rate has approached 73% of all assets being flown. Marines at all levels have contributed to this success, critical to our mission of being able to fight today.*

Major General Cederholm emphasized reshaping training and T&R manuals for future fights. While the Marine Corps maintains peer competitor capabilities, T&R manuals appropriately focused on CENTCOM operations during Middle East years. This must change.

What types of missions do we need for the evolving peer fight? How can we write T&R manuals that train to those missions, not just what we've done over the past twenty years? he asked.

2nd MAW currently has planners in the EUCOM AOR smoothing operational barriers around the European continent while eyeballing future alignment with force design priorities.

Major General Michael Cederholm gives his remarks during the 245th Marine Corps birthday cake cutting ceremony at Marine Corps Air Station Cherry Point, North Carolina, November 10, 2020. (U.S. Marine Corps photo by Lance Cpl. Yuritzy Gomez).

How do we plug into the 2030 operating concept, what tools do they need, and what missions must they train to? How do they integrate more effectively with the Joint Force and alliances? How do they integrate into the kill web? We are working on that roadmap right now, requiring significant shifts in how we educate and train formations.

F-35s are currently at 2nd MAW for training, with full receipt over the next five years. The CH-53K will arrive as the latest USMC aircraft, with VMX-1 detachment operational testing at New River.

Cederholm highlighted a critical perspective from what he called an "incredible defense leader": *Why do we stuff the F-35 into our* current operating concepts? Why don't we take our current operating concepts and revise them based on F-35 capabilities?*

Regarding the CH-53K, after recently flying the aircraft, he noted: *I was amazed at the automation built into the aircraft. I can't stop thinking about different possibilities for how this platform can support our operating concept on today's battlefield and tomorrow's.*

U.S. Marine Corps MajGen Michael S. Cederholm flies the CH-53K "King Stallion" at Marine Corps Base Camp Lejeune, North Carolina, June 12, 2021. (U.S. Marine Corps photo by Cpl. Yuritzy Gomez).

2nd MAW operates globally as "America's Air Wing." As Cederholm explained: *We operate all over the globe. Right now, we have forces from Europe to the Far East and everywhere between. The sun never sets on 2nd Marine Aircraft Wing. We must have a ready force generating combat power today while facing transformation challenges tomorrow.*

The focus remains on fully mission capable aircraft availability which is the critical metric for combat operations. *When we send aircraft into harm's way, we owe aircrew and Marine Riflemen fully mission capable aircraft*, he emphasized.

This drives intensive focus on parts and weapons systems reliability across every aircraft type at 2nd MAW. *Even with higher upfront costs for enhanced reliability, it will be cheaper long-term for operating a more resilient force and it is essential when the demand is to fight right now when the phone rings.*

The shift to address pacing threats requires operating as a distributed but integrated force. Marines must prepare to shift between supporting commands and becoming lead elements in operations. This problem-solving flexibility extends especially to assault support innovations.

Changes, great changes, in Marine Corps assault support have always originated in 2nd MAW and today is no different, Cederholm noted.

The Marines are revising logistics table of organization and manpower, seeking new balances between contracted maintenance and uniformed maintainers to free up front-line squadron capability.

2nd MAW doesn't wait for force design initiatives to increase lethality and transform operating concepts. New approaches include operating AH-1Zs and UH-1Ys with the Ground Combat Element, training Romeos with Vipers for better fleet self-defense, and focusing on Marine Air contributions to surface and subsurface missions.

A key flexibility enhancement involves rapid amphibious deck reconfiguration. *We can swap out composition on an amphibious deck within two hours to tailor force to mission or threat. We can configure for HADR operations and swap out with a ship like USS America into a full lethal strike asset with F-35s and Ospreys onboard.*

Cooperation with 2nd MEF exemplifies this mindset change. Rather than mass regimental lifts, smaller GCE elements combine with smaller ACE elements, *operating in chainsaw-like fashion*. Every seat on assault aircraft and every lifted pallet must have designed purpose for force inserts, changing resupply concepts for insertion forces.

As Major General Cederholm summarized: *We are generating combat power and transitioning simultaneously.*

2nd MAW stands *ready to answer the phone, time now, takeoff, and beat any and all potential adversaries* while retooling for future fights through innovative training, enhanced readiness, and strategic platform integration aligned with 2030 force design objectives.

The 2020 View from 2nd Marine Air Wing: The Perspective of Major General Cederholm

December 15, 2020

With the rise of 21st century authoritarian powers working skill sets for full spectrum warfare, 2nd MAW faces the challenge of being able to fight now while preparing for the future by leveraging current operations and shaping new approaches.

As a ready-to-fight-now force, 2nd MAW works with what they have while opening the aperture to rethinking how to use force elements in new ways. These changes include new ways to operate the AH-1Zs and UH-1Y with the Ground Combat Element, and new training approaches with 2nd MEF to deliver new combat approaches.

Recent exercises highlighted new ways to leverage assault support and operate in an extended battlespace. Romeos are training with Vipers to give the fleet better self-defense capabilities, with new focus on how Marine Air works with the fleet to contribute to surface and sub-surface missions.

The CG highlighted how Marines working with the Navy can enhance combat flexibility within the fleet. The shift from the ARG-MEU to the amphibious task force can allow that task force to provide significant contributions to sea control and sea denial.

We are changing our mindset. We can swap out the composition on an amphibious deck within two hours to tailor the force to the mission or the threat. We can configure for HADR operations and swap out with a ship like the USS America into a full up lethal strike asset with F-35s and Ospreys onboard. Mix and match and swapping out assets is a part of working the chess board for 21st century combat operations.

Another example of mindset change involves 2nd MEF/2nd MAW cooperation. With the pacing threat, they may not conduct mass regimental lifts. Instead, they're working on battlefield planning and training focused on correlating insertion forces with appropriate assault support.

You are taking a smaller element of the GCE, combining it with a smaller element of the ACE, and operating in a chain saw like fashion. This means that every seat on the assault aircraft, every pallet being lifted, has to have a design purpose for force inserts. We are changing the way that we think about resupply for the insertion force.

In short, the challenge is to operate now while generating change. As Major General Cederholm put it: *We are generating combat power and transitioning at the same time.*

MajGen Hedelund Underscores the Challenges of Innovation While Being a "Fight Tonight" Force

July 8, 2014

During my recent visit to the 2nd Marine Aircraft Wing, I've had the opportunity to observe the Wing's important role as a force for innovation. KC-130Js and Ospreys represent significant parts of that innovation, along with the Special Purpose MAGTF generated from here.

But innovation takes time.

My visits over the years to the Osprey squadrons illustrate this point. During one of my first visits, an Osprey pilot noted that one challenge in Iraq was convincing Marines to get off the plane because "we can't be there yet."

Later, another member of what the Marines call the "Osprey nation" observed that a sign of progress was "we are no longer a bar act." Now the Marines are the only tiltrotor-enabled assault force in the world, moving ahead with other innovations enabled by this capability.

I had a chance to meet with MajGen Hedelund and discuss the process of change with him.

U.S. Marine Corps MajGen Robert F. Hedelund, the commanding general of II Marine Expeditionary Force speaks with Marines during Exercise Trident Juncture 18 at Vaernes Air Station, Norway, Nov. 7, 2018. U.S. Marine Corps photo by Lance Cpl. Tanner Seims).

MajGen Hedelund: *If I reflect back on my own experience with the Osprey, I can certainly underscore that innovation takes time. When I was getting ready to get my wings in Pensacola in 1985, a senior Marine told me I was going to be a lucky Marine because I'd go to the East Coast CH-46s for one deployment and then to the MV-22 for the rest of my career.*

My next checkpoint was a stop at the Boeing plant in 1988 to see the first Osprey on the planet which was painted in General Gray's camouflage. There I was, Captain Hedelund, looking into a cockpit that would never be used in an operational Osprey.

In 2001, I was slated to become a VMM commander but became an HMM commander because the V-22 program was in an operational pause. When I became the CO of MAWTS-1 in 2006, I had my first crack at flying the airplane. The experience was an eye opener regarding what a key platform it could be for the USMC and how it would change amphibious assault.

Iraqi operations was really the first time that MAWTS-1 got their hands on a living, breathing V-22, tasked with integrating the plane into Iraqi operations and the weapons and tactics instructors course.

When I first came to New River several years ago, there were very few Ospreys on the tarmac. When I returned earlier this year with Murielle Delaporte, she was surprised by the number of Ospreys on the tarmac. As a frequent interviewer of the French Air Force, she pointed out that numbers matter.

MajGen Hedelund: *We are only talking about a few years, but the changes have been truly stunning. We're taking that operational experience and marrying it with innovative thinking regarding the F-35, UAVs, electronic warfare integration, and digital integration of the assault force.*

For us, innovation is blended with a combat culture that innovates for a purpose which is to succeed in difficult circumstances. With the Osprey, we're not thinking in rotorcraft terms. We're thinking in big chunks of operational space and figuring out how to operate more effectively within expanded battlespace.

When I say speed is life, I think you can do things with a relatively light force by being ahead of them in situational awareness and reach, so you can get in, accomplish something, and get out before the adversary knows you're in their backyard.

The KC-130J and Osprey pairing is changing how the USMC operates.

Another major change involves electronic warfare. We're working with UAVs, Prowlers, and the F-35 to reshape how we think about electronic warfare.

A capability like Harvest Hawk has revolutionized how we look at Nontraditional Intelligence, Surveillance, and Reconnaissance (NTISR) and delivery of precision fires for it is a game changer, simply said. Tactical electronic attack is an art form that enables thinking through how to operate a force in a contested operational area.

Question: How is 2nd MAW getting ready for the future, including the introduction of F-35s into the MAGTF?

MajGen Hedelund: *We want to accelerate the innovation the F-35 can bring to the USMC. We are digital immigrants; the operators of the F-35 are going to be the digital natives. Those natives are going to pioneer how the plane will be used. My job is to facilitate their efforts.*

Our UAV squadron is pushing the envelope on change. We're flying the RQ-21 Blackjack, currently operating it in Afghanistan to find out what it can and cannot do. We're focusing less on the air vehicle than on payload innovations. The RQ-21 enables us to work with operational capability that can be used ashore or off a ship. The critical piece is the wide spectrum of payloads it can carry.

We certainly don't want to repeat one key experience from introducing the MV-22 into the USMC. It was poorly described as a "medium lift replacement" for the CH-46. The F-35 is not a replacement for anything; it is a whole new capability for the MAGTF and needs to be approached as such from the outset.

Question: How do you view MAWTS-1's role in getting the USMC ready for the F-35?

MajGen Hedelund: *It's central. The ADT&E Division within MAWTS-1 will be hungry for operational capability of the JSF. There will be natural magnetism between those in 121 who want to kill the enemy sooner and those at MAWTS-1 who want to standardize and integrate how to do it.*

It's no accident that the CO of 121 right now is a former MAWTS-1 instructor who worked for me. LtCol Gillette was in the F-18 Division when I was the CO. It's very healthy that VMX-22, VMFA-121, and MAWTS-1 will be sharing their backyards. The smart guys and gals in those units will drive innovation like nowhere else in our Corps.

That relationship is absolutely vital to getting as much out of that squadron as we can, both before they deploy and once they do deploy.

2nd MAW Forward: The Role of Airpower in the Afghan Operation

April 16, 2012

During the visit to 2nd Marine Corps Air Wing in early March 2012, I had an opportunity to sit down with General Walters to discuss his recent experience in Afghanistan.

As commander of air operations for 2nd MAW (Forward), Walters had an opportunity to see evolving air operations support to the ground element in Afghanistan. In an earlier interview, Walters underscored the nature of air support in Afghanistan.

According to Walters: *About 18 months ago, it was all about Marjah, but now it's more about Sangin and the fight up north.*

The war has shifted. We as aviators have to adjust what we are doing to better support the fight up there.

General Walters after the interview, 2012.

In 2010, U.S. Marines and their coalition and Afghan partners carried out a joint offensive, Operation Moshtarak, to rout insurgents from Marjah in Helmand province's Nad Ali district.

But recent changes center on providing more support in volatile Sangin district. These include establishing forward arming and refueling points to maximize close-air support, creating a detachment of attack helicopters to provide cover for air ambulances, and moving Marine Corps unmanned aerial vehicles north to help in surveillance efforts.

We have operationally shaped the battle for what the conflict has presented us, said Walters.

As the battle on the ground shifts, so too must air support.

Partnership for Marines in Afghanistan isn't limited to their own air and ground forces. Since the beginning of their increased presence in southwestern Afghanistan, U.S. Marines have not fought alone.

The Marines of 2nd Marine Aircraft Wing (Forward) live work and fight side-by-side with coalition partners, particularly members of the British Armed Forces.

General Walters just arrived from Afghanistan to prepare to become the CG of 2nd MAW. And during the meeting with the current CG, MajGen "Dog" Davis, there was a chance to discuss his recent experience in Afghanistan.

Question: General Davis has mentioned the use of the Ospreys and how they are changing operations. How you use the Ospreys in Afghanistan?

General Walters: *They had their normal fair share of general support, resupplies, etc. But we started accelerating their use as my time there went on, and used them for both the conventional and Special Forces operations.*

The beauty of the speed of the Osprey is that you can get the Special Operations forces where they need to be and to augment what the conventional forces were doing and thereby take pressure off of the conventional forces. And with the SAME assets, you could make multiple trips or make multiple hits, which allowed us to shape what the Taliban was trying to do.

The Taliban has a very rudimentary but effective early warning system for counter-air. They spaced guys around their area of interest, their headquarters, etc. Then they would call in on cell or satellite phones to chat or track. It was very easy for them to track. They had names for our aircraft, like the CH-53s, which they called "Fat Cows."

But they did not talk much about the Osprey because they were so quick and lethal.

And because of its speed and range, you did not have to come on the axis that would expect. You could go around, or behind them and then zip in. We also started expanding our night operations with the Osprey. We rigged up a V-22 for battlefield illumination.

A lot of these mission sets were never designed into the V-22 but you put it into the field and configure it to do the various missions required. And we have new software for the Ospreys in Afghanistan where you can pick your approach, angle, approach speed and let the aircraft do it all. That is a huge safety gain.

Comment: And this is occurring in a very difficult environment to operate for sure.

General Walters: *In a very difficult environment. If you had to go, we would prefer to send in a V-22. You did not have to worry about that young* captain or lieutenant flying it in and losing the bubble a little bit. *You could just let the airplane do it for you. And the fact that you could get the force where it needs to be very quickly is becoming a staple of our operational approach.*

Comment: Both to get in and to get out quickly.

General Walters: *Or to get out quickly. And the speed and the agility of the aircraft also enhance survivability.*

During my year there, I only had two V-22s that got a hole in them for an enemy bullet. Every other helicopter, including the British, was hit frequently. Because of their speed, the Ospreys could get in and out and move from the threat area quickly.

Question: As we face transition in Afghanistan, one option clearly is to rely more on the Special Forces type of support to the Afghans against the insurgency. Your experience in many ways presages such an effort. How would your experience shape understanding from a professional military point of view of how to best support the Afghans with a Special Forces type of support?

General Walters: *Our role will be to support the Afghan security forces. You're going to have to support those guys, and they're going to be much more distributed. You're not going to have the battalions out there that you support people on the FABs. It's going to have to be from a central location. And the QRF (Quick Reaction Force) is going to have to be good, and it's going to have to be there quickly.*

In the end, we have to be able to prove to the Afghan security forces that if something happens, this platoon is good enough until we get someone in there.

If you ever need more than a platoon's worth of trigger pullers in a district center, the V22s is how you're going to get there quickly and decisively enough to matter.

The Afghan National Army and Afghan Security Forces understand from their perspective, how important air is. We have made them big consumers.

They know that the air is there for them; they'll go out and operate. I've had more than one brigade commander tell me that if it wasn't for the medevac, it wasn't for the resupply, and if it wasn't for the aviation fires, he didn't think he could get the battalions out operating like they do. Because they've learned that if they get hurt, we'll fix them. They know if they run out of bullets, we'll get them bullets. And if they're hungry or thirsty, we'll get them food and water.

From Afghanistan, to Bold Alligator 2012, to the Future: The USMC Re-shapes Its Capabilities

February 8, 2012

In a wide-ranging interview, MajGen "Dog" Davis, the Commanding General of 2nd Marine Air Wing, provided an overview of how USMC aviation operates now and is evolving for the future.

The bottom line for USMC Aviation was clear throughout the discussion: *We task organize the air combat element for whatever we need to do, whether it's a big fight for the Marine Expeditionary Force (MEF), a Marine Expeditionary Brigade (MEB) or a smaller but no less strategic fight for the Marine Expeditionary Unit (MEU). The most important component in that entire calculus for us is the Lance Corporal, the point in the spear*, said Major General Davis.

The Marine ACE exists for one reason, to make our Marines better fighters. When we improve the ACE's capability to fight -- we increase the warfighting effectiveness of our riflemen. At the end of the day if that kid needs support, he's going to get it. We might lose airplanes; but he's going to get support.

Davis recently visited the 2nd Marine Aircraft Wing Forward where 3,000 Marines and 85 aircraft operate in RC Southwest and at Bagram Air Force Base. What he found was both inspirational and remarkable.

We've had terrible weather over there -- dust like we had in Iraq but also low cloud, rain and ice. The real challenge has been shaping a logistics infrastructure allowing for our operations and support to combat, Davis explained.

The key to success was FOB Dwyer. *We knew when we went into Afghanistan the Marines were going to take this town called Marja. In order to take Marja, you needed to build up combat power as quickly as possible and posture your air support assets to surge support and exponentially increase the sortie rate.*

FOB Dwyer's 4,000-foot strip allowed KC-130s to flow supplies and provided runway space for hot rearm and hot refuel of AV-8Bs. *With FOB Dwyer, a 5 minute flight to be overhead Marja, we could drop in, get hot gas and re-arm without ever shutting down our motors, then launch and be* overhead Marja 5 minutes later. By investing up front in FOB Dwyer, we could take 10 STOVL attack aircraft and make 10 airplanes perform like 40 anywhere else.

The results were impressive: *The Harrier squadron that's over there, 10 airplanes flew 900 combat hours last month. They run 8 to 9 of 10 up every single day. This exceptional readiness extended across all platforms — KC-130s, V-22s, Hornets, CH-53s, and UH-1s.*

The USMC has pushed legacy equipment hard while transitioning to revolutionary new capabilities like the MV-22 Osprey. *We've extracted a heck of a lot of capability from what we call legacy airplanes. But there's a limit on how much you can push out of those things. We are fighting the legacy airplanes so very differently than the way I fought them when I was a lieutenant.*

The MV-22 represents a quantum leap: *We replaced CH-46 with a V-22. It's not a replacement. It's a totally different system, and it's changing the way that we fight, not just in Afghanistan but everywhere. It flies as fast as a KC-130, can fly low or medium altitude, and can hover like a helicopter. We've just scratched the surface on what we can do with that machine.*

The Libyan operations showcased how legacy and new systems work together. When Operation Odyssey Dawn began, a Marine Expeditionary Unit off Libya's coast had only a partial ACE -- 4 MV-22s, 3 CH-53Es, 6 Harriers and 2 UH-1Ns, since half the MEU ACE was in Afghanistan.

The first night of that operation you had this Marine amphibious ship which a lot of people said, well, it's just got these Harriers on there. But the Harriers are a vastly different machine than when I was a lieutenant. The upgraded Harriers with new avionics, sensors, and Single Seat FAC(A) capability proved crucial.

When an F-15 crew went down, the MEU ACE responded brilliantly. *The Harriers launched immediately, got up overhead the area, and established comms with the down pilot. The down pilot actually called in an airstrike from the AV-8s, targeting vehicles that were closing on his position.*

The MV-22 rescue mission was unprecedented: *From a no notice start, to launching, speeding to the pick-up zone at 250 knots, shooting the gap between several SAM sites, to landing in the LZ and getting that pilot out of harm's way at 250 knots. 45 minutes each way, instead of hours. No one could have gone and got that guy like we went and got him. Nobody.*

The coming F-35B will bring exponential improvements to MEU capabilities. *When we have the F-35B, they bring exponentially greater capabilities to our MEUs -- akin to moving from a CH-46 to a MV-22 -- and a true fifth generation capability from a small deck carrier.*

Davis emphasized the F-35's sensor capabilities: *With the F-35 we will have a superb sensor platform and the ability to connect the battle space. I think I'm going to look at it from the perspective of how much information can we extract from the F-35, and how can we share it with the MAGTF, Naval and Combatant Commanders.*

The F-35 addresses a critical gap: *One thing is we've got a glaring deficiency on board our Marine Expeditionary Units -- there's a lack of electronic warfare. We haven't had it in an airborne platform in the past. The F-35 brings that to us now.*

General Krulak's 1995 prediction of "Three Block War" scenarios has proven prescient. *You could have one block doing humanitarian operations; the middle block had peacekeeping or peace enforcement, and the third block had high intensity combat. Sometimes all three battles would rage simultaneously.*

You might take off thinking you are going to conduct an ISR mission for the 1st block humanitarian effort and end up doing danger close urban CAS in a high threat air environment on the 3rd block. A 5th generation aircraft that can see, collect, share, fight and survive will be key to MAGTF success in the 3 block wars of our future.

The evolution of USMC aviation capabilities -- from innovative use of legacy systems to revolutionary platforms like the MV-22 and F-35B -- all serves one ultimate purpose: supporting the Marine on the ground.

For us, it is the lance corporal, the point of the spear that is our focal point. The new air capabilities we are fielding and acquiring allow us to go fight in the toughest fight and survive and come back and go do it again.

"The Future is Now": Clearing the Decks for the "iPad Generation Pilots"

October 4, 2011

"Dog" Davis during the 2011 interview.

During an interview at Cherry Point Air Station, the Commanding General of the 2nd MAW discussed the evolution of the Amphibious Ready Group and its aviation capabilities.

MajGen Davis highlighted the inherent flexibility of the Amphibious Ready Group (ARG) and Marine Expeditionary Unit (MEU) structure. He described MEUs as having a "job jar" for each deployment, one that has proven dramatically different each time they set sail. Unlike rigid mission structures, the MEU's capabilities can be shaped on the fly, making them uniquely valuable to Combatant Commanders worldwide.

According to MajGen Davis: *The Marine Corps, whether we're maneuvering from forward operating bases ashore or maneuvering from sea bases afloat is an immensely valuable force. The Combatant Commanders keep asking for more of what we produce for deterrence and combat operations, in every corner of the globe. Marine Forces are so utilitarian, offer such strategic agility, and can be very readily tailored to be right sized for the task at hand that our demand signal is way up.*

The MV-22 Osprey has fundamentally transformed ARG capabilities through its superior speed, range, and operational flexibility. The aircraft operates at twice the speed and range of conventional helicopters, dramatically expanding the Marines' ability to influence operations across vast territories.

The Osprey allows the warfighter to influence operations over a vast chunk of real estate at speeds and ranges twice that of conventional helicopters; the operational and strategic significance of this capability should not be unappreciated, Davis explained.

The Osprey's survivability advantages are equally significant. Its ability to fly above primary threats, combined with its quiet operation in airplane mode and unpredictable flight patterns, makes it extremely difficult for adversaries to counter effectively.

Beyond combat operations, the Osprey has revolutionized logistics support. Recent 26th MEU operations demonstrated the aircraft's ability to sustain ARGs at extended ranges, four times greater than previous capabilities and at twice the speed.

Just two days ago we saw a MV-22 deliver a Harrier engine from the supply ship over to a big deck and that hadn't been done before, Davis noted, emphasizing how this capability allows for distributed fleet operations with unprecedented material movement efficiency.

The Osprey's strategic mobility was dramatically illustrated when aircraft flew directly from Afghanistan to rejoin an ARG, completing the journey from Afghanistan to Kuwait to Souda Bay in under 13 hours of flight time.

While praising current AV-8B Harrier improvements, MajGen Davis emphasized that the incoming F-35B represents an incomparable leap forward. The F-35B will exponentially change ARG capabilities and fundamentally alter how Combatant Commanders view MEUs and amphibious operations.

What we have coming in the F-35B as compared to an AV8-B is absolutely no comparison. This airplane is going to change the capability of the ARG exponentially in a very significant and positive direction, he stated.

The combination of Osprey and F-35B capabilities creates

powerful synergistic effects. Where adversaries might develop countermeasures against the Osprey based on its Libya operations, the F-35B's escort capabilities will expand where and how the Osprey operates.

With the F-35 escorting that airplane, it will be able to expand where the Osprey goes and under what conditions. I will be able to go after an air threat, ground threat and provide air-ground fires deep in the enemies battlespace in support of MV-22 operations and do all that from a large deck amphibious ship.

The F-35B brings organic electronic warfare capabilities to the ARG, addressing a critical capability gap. This represents a significant strategic advantage for the USN-USMC team across the entire spectrum of operations.

MajGen Davis highlighted the unique opportunity the F-35B presents for integrating three distinct Marine aviation communities: Harrier, Hornet, and Prowler pilots. Each brings different perspectives—expeditionary basing expertise from Harrier pilots, electronic warfare and high-end fight experience from Prowler aviators, and multi-role capabilities from Hornet pilots.

We are going to blend three outstanding communities. Each community has a slightly different approach to problem solving, he explained.

Perhaps most significantly, Davis emphasized that the newest generation of pilots — those he termed the "iPad generation"— will be best positioned to unlock these advanced capabilities' full potential.

I think it's going to be the new generation, the newbies that are in the training command right now that are getting ready to go fly the F35, who are going to unleash the capabilities of this jet. They will say, 'Hey, this is what the system will give me. Don't cap me; don't box me. This is what this thing can do, this is how we can best employ the machine, its agility its sensors to support the guy on the ground, our MEU Commanders and our Combatant Commanders and this is what we should do with it to make it effective.'

This generational shift represents more than technological adaptation. It embodies a fundamental reimagining of how advanced aviation capabilities can be employed in support of Marine Corps missions worldwide.

The convergence of Osprey operational experience and

incoming F-35B capabilities represents a transformational moment for Marine aviation and amphibious operations. These technologies, combined with the innovative mindset of next-generation aviators, position the Marine Corps to deliver unprecedented strategic agility and combat effectiveness to Combatant Commanders across diverse global missions.

Gamechanger: The Evolving Amphibious Ready Group

May 2, 2011

The newly equipped Amphibious Ready Group (ARG) provides a significant shaping function for the President, offering substantial flexibility and redefining the dichotomy between hard and soft power.

The USN-USMC amphibious team can provide wide-ranging options simply by being offshore, with 5th generation aircraft capability providing 360-degree situational awareness, deep visibility over air and ground space, and carrying significant capability to empower a full spectrum force as needed.

The evolution of the ARG and its impact on national security policy was discussed during a March 2011 interview with General "Dog" Davis, Commander of the 2nd Marine Aircraft Wing at Cherry Point, North Carolina.

Question: How would you describe the changes which the new aircraft provide to enhance the capabilities of the ARG?

MajGen Davis: *I would start with the impact of the Osprey. The range and speed of the Osprey create a whole different situation in terms of the radius of operational impact of the ARG.*

From the deck of an amphibious ship, MV-22s quadruple the ranges we've been able to fly at twice the speed we've been accustomed to in the past. The MV-22 changes all the equations in the Med, the Persian Gulf, and will do the same when we deploy to the Pacific.

Add in a KC-130J that can lift from a short or austere field and provide fuel for our MV-22s, AV-8s and CH-53Es, and we expand the "reach" of our MEUs exponentially.

Last month we took off with four V-22s and two C-130s to practice a self-deploy to Central America. With internal fuel loads, we could have topped them off once over Key West and flown all the way to Belize or about a five and a half hour flight from North Carolina.

What's really interesting is that during this training mission, one V-22 had a problem after takeoff and had to turn around. It landed at New River, got another airplane, took off and caught us over Gainesville, Florida. We were all doing about 200-210 miles an hour behind the tankers at 10,000 feet, and this aircraft caught up with us. You wouldn't be able to do that in a conventional helicopter. That's operational reach with potentially strategic impacts.

Question: How important will it be to get the larger ship, LPD-17, as the launch deck for the new aircraft?

MajGen Davis: *We just finished Southern Partnership Station 11 with a Special Purpose MAGTF. They did a fantastic job in Belize, but were limited by the fact they didn't have aircraft with them on their ship. The Gunston Hall's flight deck wouldn't accommodate aircraft at sea for an extended period so you could land them but couldn't sustain them.*

With the LPD-17, I've got command and control, and I've also got a flight deck where I can take bigger airplanes aboard and operate. It's absolutely key to have the LPD-17 in numbers. It's a fantastic vessel offering the nation a capability that is in very short supply compared to demand.

Question: Looking at the Libyan scenario, could you discuss how the newly enabled ARG would operate?

MajGen Davis: *Currently we have C-130s flying in Libya. If I could pair those C-130s with F-35Bs, I can provide multi-mission support and be available for other operations. The F-35 has an EW capability that no one else has. It can actually help jam and support platforms flying from point A to point B. Put a Next Gen jammer on the F-35B, and you've got very high-end EW/jamming capability for self-protection and to protect our assault support assets and ground forces.*

For many years our MEUs haven't had an aviation EW capability. With the F-35B introduction to the Fleet Marine Force, we'll have that capability, changing how we view MEUs and opening the aperture to a much wider range of missions.

You also have very high-end air defense capability with F-35Bs, VLO, fantastic radar and situational awareness, and state-of-the-art air-to-air weaponry. Add in that tethered KC-130J, and you have even greater capability.

Question: The ARG is a shaping function force. How will adversaries look at this role in the future?

MajGen Davis: *I'm Muammar Gaddafi or whoever, and I've got an ARG with this new gear embarked. I can't help but think it's going to change how I view that force. That ARG can reach out and touch me from long range, landing high-end infantry forces deep inside my territory at twice the speed anyone else can. These new capabilities will make MEUs exponentially more potent and useful to our nation's leadership.*

The F-35Bs give the new ARG a very high-end air superiority fighter that's low observable when needed. I can roll from Air-to-Air to Air-to-Ground quickly and be superior to all comers in both missions. I can use F-35s to escort V-22s deep into enemy territory. With those V-22s we can range out to a 400-500-mile radius from the ship without air refueling, delivering Marines deep in enemy territory at 250 miles an hour.

I also have significant mix-and-match capability. If I wanted to strip some V-22s off the deck to accommodate more F-35s, I could do so easily. Their long legs allow them to lily pad off a much larger array of shore FOBs while still supporting the MEU. I can quickly reconfigure from being a helicopter-centric amphib to a fast jet amphib, or conversely, take F-35s off and load up with V-22s, 53Ks, or AH-1Zs and UH-1Ys.

We'll have a very configurable, agile ship to reconfigure almost instantly based on the situation at hand. The enemy will look at the ARG as something completely different from what we have now. The newly enabled ARG will force our opponents to look at things very differently.

We will use it differently, and our opponents will view it differently.

FOUR

2009 Visit: Marine Air in the Land Wars

My visit in 2009 highlighted the USMC's adaptation of air assets for counter-insurgency operations and the introduction of transformative capabilities through the V-22 Osprey program which was first introduced into combat in Iraq in 2007.

This initial visit highlights my overall experience with 2nd MAW over the years. They were preparing to deploy or were deployed forward. They worked back home on new equipment and new techniques which they could use in the fight tonight. "Readiness" meant more than available aircraft: it meant having the right mindset for the battle after next which was just around the corner for the Marines at 2nd MAW and II MEF.

Brigadier General Walsh on Iraq Operations

November 30, 2009

General Walsh during the interview.

On November 9th, Brigadier General Walsh discussed his recent experiences commanding the 2nd Marine Aircraft Wing Forward in Iraq from December 2008 through November 2009. His extensive aviation background, including service as principal deputy to the Deputy Commandant for Aviation and as an instructor at Top Gun, provides unique insight into airpower's evolving role in counterinsurgency operations.

During Walsh's deployment, two critical transitions were underway: the acceleration of stability operations and growing collaboration with Iraqi forces developing their own security capabilities. This created two crucial tasks for USMC aviation:

- Supporting U.S. forces as they transitioned to advisory roles.
- Assisting Iraqis in developing independent security operations.

The core role of USMC air became largely non-kinetic, focused on presence and support rather than strike missions. Walsh emphasized that the shift is from shaping air around precision-strike to shaping air to provide collaborative presence.

He underscored a key challenge: *One can measure the effects of kinetic strike; it is more difficult to measure the effects of presence.*

Walsh described the fundamental challenge facing COIN forces:

You are not dealing with one large formation on attack; the forces are very decentralized and very distributed. Operating across Anbar province — 250 miles by 150 miles — with companies and platoons scattered across vast territory meant supporting dispersed forces with limited assets in an environment with no rear area or safe zones.

Walsh noted: *The enemy can hit you anywhere; the enemy gets a vote, We are not on the offensive; we are on the defensive. Both the enemy and we are living among the people, and the challenge is to get them on our side.*

This created a central question: How does aviation provide support in such a chaotic environment where both ground forces and pilots must manage constant uncertainty?

Walsh identified three major levels where air presence proved significant:

First, supporting Marines on the ground. This included lift operations, overwatch, fire support capability, low-level flights to demonstrate available firepower to local populations, and response to dispersed enemy attacks. Walsh characterized this as "no Marine walks alone" or when Marines operate "outside of the wire," airpower provides protection and support.

He gave an example of IED response: *When you have a vehicle blown with an IED and have the road divided into slow-moving lanes, how do you know who is in those vehicles? A request would come in: Please bring in a fighter for presence to show you are there. How do you measure that effect?*

Second, reassuring the population. As Iraqi leadership assumed more functions, air presence helped reassure populations that support remained available. When Al Anbar's provincial government was seated in June 2009 amid Al Qaeda threats to Ramadi, Walsh underscored: *The Governor asked us to fly our F-18s at 5000 feet to reassure the population and to deter any threats.*

Third, deterring attacks from dispersed adversaries. Walsh described being at a checkpoint under mortar fire: *We called in some F-18s and the minute the planes showed up the firing stopped; the enemy figured out that the F18s would know where they were with the obvious consequences. How do you measure this effect?*

Paradoxically, as U.S. forces withdrew, demand for airpower increased rather than decreased. Combat posts were reduced from

140 to just 4 during Walsh's tenure, meaning "air was relied on more frequently for convoy protection" and to provide capabilities previously handled by ground forces.

Additionally, retrograde operations required extensive air support for equipment movement and increased transport needs for partnering operations with Iraqi forces.

COIN operations required fundamentally different concept of operations (con-ops) planning compared to strike-oriented missions. Instead of advance tasking with identified targets, presence missions required on-the-fly planning with ground and air elements using common mission planning software.

Pilots would launch with 4-5 potential taskings and shift among them based on ground force demands. Ground forces would identify their daily schedules and intentions through mission software and web-based communication systems, driving air element taskings.

Walsh commented: *Rather than carrying ordinance to cover a target, we are going airborne to support demands shaped by the most likely cluster of activities,* Walsh explained.

The traditional C4ISR approach for precision strike was replaced by a collaborative decision-making model emphasizing shared situational awareness between ground and air elements in real-time. A "push concept" was adopted where air elements launched in support of ground forces, selecting from pre-planned presence support lists while adjusting for emerging challenges. This differed from the USAF approach of having airpower ready with weapons for on-demand strikes.

Blue Force Tracking (BFT) became crucial technology, providing real-time location data and text communication capabilities. Walsh was adamant about its importance: *I can do my operations with real knowledge of where the ground elements are located to support them,* Walsh noted. *All of our fixed wing aircraft should have BFT on them and I hope F-35 will have as well.*

The Direct Air Support Center (DASC) evolved from an air traffic control function to a battle management role, allocating resources against fluid demand structures in what Walsh called "distributed chaos." The USMC's institutional investment in air-ground integration proved essential for COIN operations.

Unlike the USAF's limited presence within Army units, the USMC maintains significant aviator presence integrated with ground elements down to the company level. This is how he put it: *When the 82nd Airborne replaced the USMC force, the USAF had one captain as an air liaison officer; we have 12 for every one captain that the USAF has,* Walsh observed.

This integration will be enhanced by future platforms like the F-35: *The processing power of the aircraft and the software on board which will allow us to support directly overhead our ground forces will be an exponential increase in capability. But this capability will be built upon the organizational investment and the habitual relationship we have between the ground and air elements.*

Central to USMC effectiveness is unified command with distributed control. While command can be centralized, control cannot be in COIN environments, requiring air elements to provide support through significant ground-air integration and on-the-fly planning.

Walsh's insights reveal how counterinsurgency operations fundamentally transformed airpower employment, shifting from precision strike to collaborative presence. This is a change with significant implications for future military operations and inter-service cooperation.

The Maintainers at the Heart of the Osprey Enterprise

November 28, 2009

The Osprey is a new plane, a new capability, and provides a new contribution to the MAGTF and to the joint forces. It is not simply a replacement for the CH-47 or CH-53. As such, being a new platform and a new capability, the Osprey has challenged USMC thinking about concepts of operation and its impact on joint operations has not really been understood.

As a senior member of the Obama Pentagon commented when asked "have you thought about the impact of tiltrotor technology on con-ops," he responded simply no. Yet the Osprey is one of the truly innovative capabilities being introduced in the USMC and the joint forces.

The V22 Maintenance Team

Not surprisingly, being new requires new approaches to supporting the aircraft as well. It does not perform like a rotorcraft, so it will not be maintained with approaches inherited from USCM experience with rotorcraft. In effect, the aircraft with on board sensors, and data flows available to both the pilot and the ground crew calls for a whole new approach to maintenance.

In effect, an Osprey enterprise is being forged in which the pilots, maintenance crews and the contractor support teams are working together to shape effective approaches to support the aircraft. The maintainers are not the last members of the team, but integral players in shaping the evolving approach to sustainment of the aircraft.

This perspective was indeed clearly conveyed in an interview with members of the Osprey maintenance team for the Osprey squadron VM-266 currently stationed in New River air station in North Carolina

In the conversation with the maintenance crew, the fundamental differences with maintaining the CH-46 were underscored: *This is a whole new bird,* commented one member of the team. The "oldest" member of the team had five years experience with the Osprey and the team underscored that the core effort to maintain the aircraft was being done by younger Marines for whom the Osprey is their first aircraft.

This means that the teachers of maintenance for the Osprey have experience going back only a few years; whereas, as one maintainer added: *The CH-46 was being maintained by Marines being taught by instructors with 10-20 years of experience and 3-4 deployments under their belt,* commented one Marine. Fresh approaches were required to deal with what in many ways is a revolutionary aircraft.

Training on computers and in dealing with data generated by sensors on the aircraft is central to the technical skills required. As on maintainer emphasized: *The CH-46 unlike the Osprey is not a sensor rich aircraft and does not generate maintenance data required for the support of the aircraft.*

The differences with traditional rotorcraft were highlighted throughout. As one team member noted: *The experience of maintaining the Osprey is more like dealing with the F-18, than it is like dealing with rotorcraft. The modern systems of the F-18 are more like the Osprey than a helicopter.*

Some of the members interviewed had had experience with the first deployment to Iraq of the Osprey. *We were well trained but did not know fully what to expect. We had learned how to maintain the aircraft in the U.S., but in the field it is a whole different experience.*

And the team member added: *We learned a great deal while on deployment.* In fact, one Osprey observer has underscored that *we catalogued what normally would be 5-6 years of experience into 18 months.*

This observation was echoed throughout the interview. A major emphasis of several of the team members was the evolving nature of the enterprise. as noted: *We are working closely with the contractor in shaping new approaches to managing maintenance events. We send the maintenance events to the contractor through an IT system, and the contractor's engineers work through solutions which are then translated into new approaches and new maintenance standards,* noted one Marine.

MV-22Bs seen on the tarmac during the visit.

Another added, *We started with a small book of maintenance standards which has grown into a large book as we are identifying the standards which we will follow in maintaining the aircraft.*

The Marines underscored that the contractor deploys with them in Afghanistan and this working relationship was central to the dialogue between engineers, pilots and maintainers seen as crucial to shaping the Osprey enterprise. *We see progress all the time*, noted one Marine. *As Marines get to know the aircraft, they see how it saves lives and extends the battlespace. We win converts to the aircraft as Marines see it perform. We view our task as part of an overall team effort to ensure that this battlefield advantage be available to our fellow Marines.*

A core capability driving the entire maintenance effort is that the avionics drive the system. As one Marine commented: *the system provides us with the data with regard to what are the problems which need to be fixed. The challenge is to find the underlying cause to which the data indicates the problem. As we gain improved understanding of the aircraft, we are getting better at finding the causes, and then shaping more effective maintenance standards.*

In its simplest terms, the maintenance approach can be broken into three key elements.

- The first element is the role of the sensors, the mission systems and the data storage system aboard the aircraft. The sensors generate the data which is analyzed onboard the aircraft in flight and stored real time in a "brick" which contains a PC card. According to Sgt. Mireles: *There are many sensors on the aircraft. A lot of these sensors are analog sensors which send data through data convertors to the mission computers where the data can be pulled up for the pilot's use and to see where the fix can be made.*
- The second element is the processing of the data on the ground. The maintenance crew uses a COTS laptop to run the software, which is updated, on a regular basis with the contractor, and the PC card is put into the computer to run the maintenance evaluations and tests. And the exchange of data with the engineers is a central aspect of determining the evolution of maintenance standards and shaping *maintenance events*. According to Sgt. Gilbertson, *All the data is collected in the maintenance brick for the maintenance download. Inside the brick is a card where your data is stored. From here, it can be used on the (laptop) computer. From there, the aircraft summary of what the aircraft did… (can be analyzed).*
- The third key element is the dialogue between the air and ground crews. Because both teams have access to the same data, they can shape a common understanding of the problems and shape paths to resolving the problems. And Sgt. Dean underscored: *When the pilot comes in to discuss a problem with the aircraft, the way we can visualize it with him and see it with him as he does onboard the aircraft is by using the (laptop) computer where we can see the same data as he did on the aircraft.* This ability to share data is crucial for

both pilot learning and for the maintainers to understand how to improve performance of the aircraft. This ability to share data is an essential difference from legacy products.

The maintenance team is really at the heart of the evolving Osprey enterprise. They are not the last cog in the wheel of the performance of the aircraft, but are an integral part for determining the capability of the aircraft to exercise its unique capabilities in performance of USMC missions.

And the new capabilities to be maintained can lead to mission success as well. As one Marine commented: *When a CH-46 has a problem a red light comes on. You have to land it and take care of the problem. If you are doing this in Taliban country, the guy who is coming out may not have a wrench. With the V-22, not only will you know what is the problem, but you have a clear idea how long you can continue to fly with the problem.*

In other words, new capabilities provide new ways to operate, and new ways to stay alive in combat situations.

Going to Afghanistan: The Osprey Squadron Prepares

I interviewed several members of the Osprey squadron just prior to their deployment from North Carolina to Afghanistan in November 2009. The Osprey team discussed their preparation, some of their expectations and some of their thinking about how the Osprey could be used to benefit the MAGTF and the joint and allied forces.

The core view was that the unique capabilities of this aircraft would provide some tools which the MAGTF commander would be able to use in the context of the topography of Afghanistan and the demands of shifting strategy.

The speed and range of the aircraft and its ability to support MAGTF team members widely dispersed in Afghanistan were often cited in the interview. Also, underscored was the ability of the Osprey to fly at a much greater height than traditional rotorcraft to which it is often compared, and to be able to use "vertical sanctuary" by operating at higher altitudes.

Climb out in Afghanistan, 2010. Credit: LtCol Bianca.

The Afghan deployment is the second for the Osprey. The Osprey was first used in Iraq, and its "baptism" has been drawn upon for lessons learned in preparing for the new deployment. When asked whether the Marines had drawn such lessons, one squadron member commented: *Absolutely, we're the Marine Corps. We pass it on.*

Specifically, the Marines underscored that some members had previous Osprey experience in Iraq, many had Iraq combat experience and some had Afghan combat experience. Also emphasized was the wide-range of backgrounds of members of the squadron in working air issues within the MAGTF. As the squad leader commented: *We have guys from every background: Hornets, Prowlers, CH-46s, CH-53s, you name it.*

With regard to operations, the Marines discussed the challenge of getting folks to understand the impact of the new machine on operations. A common point was that: *We are not a rotorcraft; we are operating a tiltrotor craft. As such, we can do the operations of a CH-46 but we are not simply a CH-46. Do not confuse our abilities to mimic the CH-46 with the much more limited capabilities of the CH-46 when compared to the Osprey.*

As one Marine put it: *You can call it a rotorcraft, it's a form of rotorcraft, but it's a tiltrotor; that's the distinction that gives you the speed and the altitude that a normal rotorcraft doesn't have.*

Another Marine underscored that getting folks to understand the difference is essential to understanding how to use it differently from a rotorcraft: *Let me give you a practical example in CONUS. When we land in the DC area, we challenge the FAA controllers to understand how we operate. We surprised the Washington Terminal area controllers because…why can't I get in the pattern with that guy up there? I'm moving at the same speed. Because then you have to take the runway. Well, no, I don't need to take the runway, I can get off and fly helo route, too, if you want me to do that. And that just blew their minds, and we couldn't find a way to work that out. So it is not just us, the military, that are challenged to understand the unique characteristics of the Osprey, it's FAA controllers as well. We're going to have to figure out this tiltrotor piece because it is far more versatile and gives you a lot more options, and there are no rules written for it.*

Bringing the unique qualities of the Osprey to the fight is especially important in Afghanistan. This is true for several reasons.

- First, the adversary has decades of experience of tracking and combating rotorcraft. The CONOPS of the Osprey are different and can provide a counter to the years of experience of the adversary in countering rotorcraft.
- Second, the combat operations in Afghanistan will be able to draw upon the unique capabilities of the Osprey.
- Third, the ability of the Osprey to support ground forces not near its base will be a significant advantage. The aircraft can move in areas not covered by traditional rotorcraft without using multiple forward operating bases (FOB), and moving in the directions of relatively direct flight which rotorcraft need to use.

As one Marine commented on the critics of Osprey: *They simply do not take into account the operational advantages of the ability to skip a refueling or its ability to carry it all in one load.*

Another Marine emphasized the joint impact of enhanced security for the force and increased ability to surprise the enemy. *If you're flying a helicopter, you pretty much have to take a straight line in a lot of situations, but we can come from any direction, which offers a big surprise factor.*

The range of the aircraft means you can cover the entire theater. When VIPs came to Iraq and wanted to get around Iraq in a day, one Marine underscored: *The minute they got there and everybody realized that you can cap all six FOBs in less than six hours if you've got the speed and the legs to do it, the next thing you know you've become the VIP platform. Why? Because I have to get to these places before the sun sets today. And no other machine can do that for you except the V-22.*

The infrastructure piece is key to the Osprey advantage. The Osprey can operate from a single base, but its ability to operate all over the area of operations (AOR) means that it can go where it is needed.

As one Marine put it: *We're not married to the base and ground infrastructure the same way as traditional aircraft and the mission in Afghanistan requires that that not be the case. You can't do it. You couldn't or you wouldn't be able effectively to maintain aircraft and maintain the maintenance or the operation of infrastructure for a relatively small air element in so many different locations and FOBs from the company level, in some cases down to platoon level.*

But if you put them all in one place with their ability to quickly dash out and get to that guy and do whatever he needs you to do, then return to that same central base, we're in effect doing distributed operations.

The ability of the Osprey to move rapidly to support dispersed forces is central to tempo-setting. As one Marine noted: *The mission of support, which people often lose sight of, isn't just to move things around the battlefield in a circulatory motion like we've seen in Iraq, it is the ability to provide mobility to the MAGTF Commander anywhere, anytime, anyplace, any payload; that's the key. I can be wherever the enemy is, and I can be there faster than the enemy can respond to me. That's tempo-generating. That's basic maneuver warfare. I can move faster, farther, with more stuff than you can, and you can't get away from me; and you can't catch me.*

Daylight Assault in Afghanistan 2010. Credit Photo: LtCol Bianca.

The ability to quickly move ground forces from one area where they are not needed to reinforce in areas under attack is essential in the Afghan theater. One Marine underscored the significance of the Osprey contribution to this CONOPS capability: *I see V-22 as a real combat multiplier with its ability to reinforce ground forces over great distances. So right now, in the traditional deal that we're in where we have platoon-sized elements spread over hundreds of miles, it's likely to be very quiet in one area, and there will be a tic in another area.*

At the ground commander's request, we could take troops from a regionally quiet area to an area where something's going on, and that's a real combat multiplier, the ability to do that with the speed that of which the Osprey is capable. That is the crux of it all. We can reinforce, cover great distances in very short periods of time, and then return those troops to their base, which might be 200 miles away, at the conclusion of an operation.

Another key aspect is the ability to fly higher and quicker as a means of providing enhanced security and greater capability to execute envelopment operations.

As one Marine encapsulated the Osprey advantage: *Obviously Afghanistan's got terrain that we're all familiar with. Osprey's got the highest altitude capability of any vertical lift aircraft. It's the only vertical lift aircraft that has an oxygen system onboard.*

Our ability to fly more than 20,000 feet does a lot for us. Obviously, we can't carry passengers at our highest altitudes, but we can carry cargo. We can retrieve passengers and we can carry passengers at lower altitudes. But flying at higher altitudes makes you a whole lot faster. I see that glossed over when this airplane is briefed.

The average ground person, or someone who's not a pilot, or even a rotary pilot, may not fully understand it. At higher altitudes, you're about a hundred knots faster than you are on the surface, in any airplane, tiltrotor or otherwise. The ability to go higher definitely makes you faster, too.

You get vertical sanctuary. The ability of the enemy to shoot you or channelize you through some terrain is reduced. So you may have a helicopter that can fly at 13,000 or 14,000 feet, a very powerful lightly loaded helicopter; but, he's still going to have to fly through passes where the enemy could establish a threat system.

Our ability to fly at more than 20,000 feet empty, or 13,000 feet when full of people, gives us the ability to fly in straight lines from Point A to Point B without having to go around mountain ranges in certain cases, and gives us vertical sanctuary and speed while doing so. That's often not captured in discussions about the airplane.

FIVE

2010 Visit to New River: An Osprey Update

In 2010, I visited Marine Corps Air Station New River where I conducted a series of interviews with Marine Aircraft Group 26 (MAG-26) to assess the progress of the Osprey V-22 deployment. These interviews captured the historical perspective of the Osprey's first three years of operational deployment.

I wrote a report in 2010 which provided a detailed treatment of these interviews including photos of the Marines interviewed. I included that report in my second tiltrotor book published in the summer of 2025 which is entitled, *A Tiltrotor Perspective: Exploring the Experience.* If you wish to read the complete narrative, you can read it in that book.

Here I am providing a summary of what I learned during the 2010 visit to New River.

What I found was that the Osprey had already fundamentally changed Marine operations by shrinking battlespace and enabling new tactical possibilities. As Captain Dwyer, an Osprey pilot who deployed aboard the USS Nassau, explained: *The speed of the Osprey allows it to operate effectively with the Harrier. And we actually split the MEU, the entire MEU, which I don't believe, had been done before in specific type model series.*

Major Lee York, a former CH-46 pilot who transitioned to the Osprey, witnessed this speed advantage firsthand in Iraq: *We took some soldiers out to the West of Iraq. The crew chief comes up to us and tells us that the guys won't get out of the plane... They said we're not there yet... The last time we did this flight it took an hour and a half. We've only been in the plane for 40 minutes.*

Performance-Based Logistics Success

The engine PBL contract emerged as a notable success story. Matthew "Digger" Howard from the USMC Department of Aviation emphasized its effectiveness: *From the Marines perspective, the guys on the flight line, we've never waited for an engine; we've never had a bare firewall in the V-22 community. That's a significant statement to be able to make.*

The `digital diagnostic capabilities proved crucial for refining these contracts. As Howard noted: *Because of the diagnostics resident in the airplane... it does take its own temperature, monitor its own vitals. It does this continuously throughout the time that the thing is operating. There is nothing you don't know about what the airplane did when it gets back.*

Building "Osprey Nation"

Captain Paul Smith described the creation of concentrated Osprey expertise: *As we come onto the neophyte type phase of operating this aircraft, we are shaping an 'Osprey Nation'... Shared assets, shared experiences are building our domain knowledge.*

This concentration of expertise enabled more efficient training and maintenance support across squadrons operating under the same MAG structure.

Operational Versatility

The aircraft's range and altitude capabilities proved invaluable in Afghanistan. Lieutenant-Colonel Garcia explained: *What the aircraft provides that the helicopter cannot is the ability to go high. We typically flew between 9,000 and 10,000 feet to get away from the ground threats; whereas, most of the helicopters flying were lower than that.*

During Haiti relief operations, the Osprey's range enabled unique operational flexibility. Captain Dwyer recalled: *Because we had the legs, we parked it about 75 miles away.. This meant that we could range the entire island and still on an every other day basis, we could range Guantanamo Bay for just a general logistics run.*

Maintenance Learning Curve

The transition from mechanical to digital maintenance systems required significant adaptation. Master Sergeant Jeremy Kirk, an experienced CH-46 maintainer, acknowledged the challenges: *One of the biggest differences is the lack of experienced maintainers for the V-22. With the Phrog being around for so long we have decades of experience. We are still learning the V-22.*

Component reliability emerged as a key issue. Lieutenant-Colonel Garcia identified the core challenge: *We have components that are supposed to last in excess of 5,000 hours, which we're routinely replacing less than a thousand hours.*

However, progress was evident. Kirk observed: *I can see significant improvements from '05 to now from just the experience levels of maintainers learning their tasks and learning the tricks of the trade on the new aircraft.*

Battle Damage and Survivability

Combat experience revealed the aircraft's resilience. Garcia noted: *We had the misfortune of receiving some damage. However, we quickly found out the aircraft is very survivable. The composites performed exceedingly well.*

Looking Forward

The Osprey's pairing with fast jets pointed toward future operational concepts. Captain Dwyer envisioned the potential: *I saw so much potential for the short take-off vertical landing attack aircraft, fixed-wing aircraft and the V-22 working together. In the future, I would have those two, the V-22 and F-35 working very closely together.*

The interviews revealed an aircraft transforming Marine operations while its support systems matured through operational experience. As Kirk summarized the maintainer perspective: *Honestly, I like this aircraft. We just needed to have the aircraft deployed actually to learn how we were going to employ it, and how we're going to maintain it. But it does take time.*

SIX

2011 Visit to New River: USMC Con-Ops in Evolution

During my 2011 visit to New River, I focused on the evolving con-ops leveraging the Osprey. The two articles from September 2011 detailed the MV-22 Osprey's crucial role during Operation Odyssey Dawn, the 2011 U.S. military intervention in Libya.

The articles contain interviews with two Marine Corps officers: LtCol Boniface, who served as XO for the Aviation Combat Element of the 26th Marine Expeditionary Unit (MEU), and Major Debardeleben, who piloted an Osprey during a critical rescue mission over Libya.

The Role of the Osprey in Operation Odyssey Dawn

September 22, 2011

During my visit to New River to get caught up on Osprey developments, I sat down with LtCol Boniface who was the XO for the Aviation Combat Element (ACE) for the 26th MEU.

The conversation focused upon the aviation assets used in the operation off of the MEU in the Libyan operation and notably the Osprey.

LtCol Boniface has extensive experience, having done many deployments on a MEU. He started with the CH-46 and has been flying Ospreys for the last three years. He is now Commanding Officer, Marine Medium Tiltrotor Squadron 266, MCAS New River.

LtCol Boniface during the interview August, 22 2011.

Question: Some folks consider the Osprey as the replacement for the CH-46 much like some folks consider the F-35B as a replacement for the Harrier. But I believe that this actually distorts the discussion because bringing both planes to the MEU is a game-changer. What are your thoughts about the process of transition?

LtCol Boniface: *The Osprey is clearly not a CH-46. It is an aircraft that can fly like an airplane, but land and takeoff like a helicopter, but it is not defined by its essential ability to operate as a rotorcraft.*

In comparison to the legacy CH-46E, I can carry twice as much, go twice as fast and go twice as long. Actually, in many cases I can almost triple the capability. Finally, it is a more reliable aircraft and ultimately a safer aircraft.

It provides the ability for the PHIBRON/MEU to keep the ARG farther out to sea if needed, thus increasing the element of surprise and keeping us safer too. With the CH-46E, you are typically operating 25-50 nautical miles (NM) from shore. As of today, I can operate 250 NM or greater from shore and I can close this 250 NMs from ARG shipping in just about an hour with 7,000 pounds of Marines, or cargo in the back.

I can actually launch at approximately 52,000 pounds due to the MV-22's ability to perform a rolling short takeoff (STO) from the flight deck as opposed to the vertical takeoff required by a traditional helicopter.

Bottom line, I can carry more Marines, cargo and equipment and I can close an objective area twice as fast while staying outside of most enemy weapon engagement zones.

It is a completely different animal. It is a true game-changer.

Comment: The Osprey is significant in logistics support for the fleet as well which was demonstrated during recent operations off of Libya.

LtCol Boniface: *I need to be very clear, the Osprey is not only supporting the USMC, but also supporting the USN-USMC team. During this deployment the USS Kearsarge suffered a mechanical loss of a propulsion screw.*

We were only able to do four knots through the water; at that point we were 300 miles from land. The only thing we could do was to get tech reps and parts out to the ship to allow us to make a best speed of 11 knots to get back into the fight. Remember you can't really launch aircraft if you can't make the correct wind across a flight deck, and you can't effectively launch Harriers to continue their strike mission with four knots of wind either.

The Osprey was the only bird we could use to close this gap, fix the ship and continue the mission we were executing off the coast of Libya.

The V22 is like driving a Cadillac Escalade compared to your dad's old truck. I'm traveling better and I'm able to carry more. It's more reliable, and efficient. It's smoother. It's safer.

Really comparing the CH-46E to the Osprey is like comparing apples and oranges. With the CH-46E, I am typically flying 300 feet at 110 knots. With the Osprey, I can be at 13,000 feet and flying at 250 knots, all with Marines and equipment in the back. I am flying an airplane, not a helicopter.

Question: When we were talking earlier, you emphasized that the Osprey's capabilities compared to the Phrog was game-changing in character. Could you elaborate?

LtCol Boniface: *It completely changes the game for the ARG/MEU and it changes the game for how the Marine Corps does business. I didn't fully real-*

ize, nor appreciate this until I was operating in some of these locations during our deployment.

Once we got into the Med for the Libyan operations during Operation ODESSEY DAWN, Naval Air Station Sigonella was our only forward support base. The Osprey functioned as a force multiplier in these circumstances. I could fly 300 miles plus from the USS Kearsarge to Naval Air Station Sigonella, land, get a quick hit of gas if needed, put five, six, seven thousand pounds of gear, equipment, troops, parts, and be back quickly to the ship within 2.5 hours.

Half of our MV-22s were conducting combat operations in Afghanistan while we were conducting combat operations off the coast of Libya aboard the USS Kearsarge. So you can do the math: half of the Osprey's conducting combat operation in Afghanistan and the other half performing combat resupply, and TRAP operations off the coast of Libya.

I wouldn't have even fathomed this expeditionary and amphibious capability 10 years ago. Also, the Ospreys from Afghanistan flew directly to Souda Bay, Crete and then onto Naval Air Station Signalla, Italy. This trip is a 3500 NM transit. This has been the longest in our short history, and they did it in one day. You can't even begin to argue or compare and contrast these facts with the CH-46E.

Question: How important to the operation was it to break the CH-46 tether?

LtCol Boniface: *A complete transformation to how we are doing business has been involved. In order for the USS Kearsarge, the ARG and the 26th MEU to stay in their operational box during Operation ODESSEY DAWN, and enable the Harriers to continue their strike mission, we were reliant on other assets to supply us. For many supply items, the Osprey provided the logistical link to allow the ARG to stay on station and not have to move towards at sea re-supply points and meet re-supply ships.*

Without the Osprey you would have to pull the USS Kearsarge out of its operational box and send it somewhere where it can get close enough to land or get close enough to resupply ships to actually do the replenishment at sea. Or you would be forced to remain where you are at and increase the time you're going to wait for this part by three, four days or even a week.

The ARG ships are only moving at 14-15 knots. At best, let's just say they move an average of 13 knots per hour, and add that up for the 300 miles that you have to sail. Now you're looking at least a day to get the needed folks, parts or equipment and then the transit time back to the operational box. The V22 will do that in a couple hours and allow the ARG/MEU to keep executing its mission.

Question: So part of your game-changing argument is something people do not usually focus on, the use of the Osprey as a backbone of rapid resupply and logistics support?

LtCol Boniface: *It was a key performance element. It facilitated the ARG's strike missions over Libya by allowing the Osprey to perform a combat resupply mission. And it did the other items crucial to the operations such as the TRAP mission and bringing in supplies to keep the Harriers operating. By keeping us within a suitable striking distance, the Osprey helped reduce the AV8-B pilots overall fatigue factor, increase that safety margin and ultimately kill bad guys.*

And if the collapse had been more rapid, and we would have needed to put Marines on the ground (for example, a possible humanitarian assistance, disaster relief mission), there is no question that the Osprey would have been central to that effort as well.

The Libya TRAP Mission: Marines' First Combat Rescue with the Osprey

In the early morning hours of March 22, 2011, the U.S. Marine Corps executed one of its most challenging and significant rescue operations in recent memory. Deep in the Libyan desert, 130 nautical miles from the safety of the USS Kearsarge, Marines conducted a Tactical Recovery of Aircraft and Personnel (TRAP) mission that would not only save the life of a downed Air Force pilot but also demonstrate the revolutionary combat capabilities of the MV-22 Osprey tiltrotor aircraft.

The crisis began at approximately 11:33 p.m. local time on March 21, 2011, when a U.S. Air Force F-15E Strike Eagle experienced equipment malfunction and crashed about 25 miles southwest of Benghazi during Operation Odyssey Dawn, the coalition effort to enforce UN Security Council Resolution 1973 in Libya. The aircraft, tail number 91-0304 from RAF Lakenheath, was operating out of Aviano Air Base in Italy when mechanical problems forced both crew members to eject at 22,000 feet. The pilot and weapons systems officer separated during their descent, landing miles apart in hostile territory as Libyan forces loyal to Muammar Gaddafi controlled much of the surrounding area.[1]

Major Kenneth Harney, the F-15 pilot, found himself alone in the desert with enemy forces actively hunting for him. The weapons systems officer, Captain Tyler Stark, was quickly recovered by friendly rebel sympathizers who later turned him over to coalition forces. Harney, however, faced a far more perilous situation as he attempted to evade capture while communicating with overhead aircraft via survival radio.

Aboard the USS Kearsarge, Marines of the 26th Marine Expeditionary Unit (MEU) had been conducting routine operations in support of the Libya campaign when word filtered in about the potential aircraft loss. Major J. Eric Grunke, a Harrier pilot with the MEU, recalled the moment: "We were preparing for another armed reconnaissance mission where we would go out and look for targets. Word started to filter in that, potentially, an F-15 had crashed. We weren't sure why, whether it was enemy air fire or a malfunction or what, so we started to determine, okay we're going to have to launch the TRAP package."[2]

The Marine Corps response was swift and comprehensive. At 12:50 a.m., AV-8B Harriers launched from the USS Kearsarge to provide immediate support for the downed pilot. Just five minutes later, U.S. military leadership approved the full TRAP mission. The rescue package assembled was two MV-22 Osprey tiltrotor aircraft, two CH-53E Super Stallion helicopters carrying quick reaction force personnel, two AV-8B Harrier attack jets for close air support, and a KC-130J Hercules for aerial refueling.

By 1:20 a.m., the Harriers were overhead the downed pilot's position, and a nearby F-16 had established radio communication with him. What the rescue forces heard painted a terrifying picture. Major Grunke, who took over as on-scene commander, later described the communications: "I just start listening to gain an idea of what's going on down there, and I can hear him, wind rustling and him whispering into his radio. At that point it all became real to me, listening to the guy whispering on the radio. This is no longer North Carolina, this is no longer practice – that's really a guy down there scared for his life."[3]

The pilot reported that five to six tactical vehicles with searchlights were pursuing him across the desert. He could hear barking dogs and gunfire as enemy forces closed in on his position. The situation reached its most desperate point when, according to Grunke, "He comes up and actually crying on the radio he says, 'tell my wife I love her.'"

Grunke's AV-8B Harrier was equipped with two 500-pound laser-guided bombs, and deadly force had been authorized to protect the downed pilot. Within five minutes of arriving on station, Grunke had identified the pursuing vehicles through his targeting pod. "I tell the pilot, 'Okay, I can see the guys ... I've got two 500-pound bombs, do you need them?' He says, 'Yes, yes I do,'" Grunke recalled.

The Harrier pilot successfully engaged two vehicles pursuing the airman, dropping both bombs with direct hits that finally convinced the remaining enemy forces to break off their pursuit. These strikes, delivered at 1:33 a.m., cleared the immediate threat and allowed the rescue operation to proceed.

Meanwhile, the two MV-22 Ospreys had launched from the Kearsarge at 1:33 a.m., piloted by Marine Captains Erik "Brillo" Kolle and Joe "Angry" Andrejack. For both pilots, this would be their first combat rescue operation. The mission highlighted the unique capabilities that made the Osprey revolutionary for such operations.[4]

Locating the pilot required coordination between multiple aircraft. The first Osprey didn't initially spot him, but Kolle and Andrejack saw a flare on the ground that led them to his position. An aircraft overhead was able to shine a laser near the downed pilot, revealing him hiding in small desert shrubs.

When the Osprey finally approached for landing, the relief was immediate and visible. Even before the aircraft touched down, the pilot "bolted toward the Osprey aircraft with his hands above his head so no one would think he was hostile."¹⁶ The pickup was so rapid that, as one observer noted, "The Marines barely even had the chance to get out."

At 2:38 a.m., one of the Ospreys had successfully recovered the downed pilot. By 3:00 a.m., the aircraft carrying him had landed safely on the deck of the USS Kearsarge. The entire operation, from launch to recovery, had taken just 90 minutes.

The Libya TRAP mission proved to be a watershed moment for the Osprey program and Marine Corps operations. The MV-22 executed the mission at least 45 minutes faster than the next available platform, demonstrating capabilities that fundamentally changed how the Marine Corps could operate. The speed and range advantages allowed the Marines to operate from a much greater distance than traditional helicopter-based rescue operations would permit.

The Libya TRAP mission stands as a testament to Marine Corps training, innovation, and courage under fire. In 90 minutes of precise execution, the Marines not only saved the life of a fellow service member but also demonstrated the combat value of one of the military's most controversial aircraft. The mission validated the Osprey's revolutionary capabilities and proved that the Marine Corps' investment in tiltrotor technology would fundamentally transform their ability to project power and conduct rescue operations across vast distances.

The Execution of the TRAP Mission Over Libya

September 25, 2011

During a visit to New River to discuss Osprey operations, I had the opportunity to speak with the ACE commander and one of the Osprey operators involved in the TRAP mission over Libya. This is his first-hand account of the events:

After we took off, the autopilot took over. That's one of the great things about this aircraft versus the SEA KNIGHT—it can fly itself through part of the operation. Like setting cruise control in a car, you can focus your mind elsewhere. You're monitoring the flying, but now you can manage the mission better.

We focused on navigating to the moving recovery zone, comparing our mission plan to the unfolding operation. This gave us time to assess everything, communicate on the radio, and build situational awareness.

Major Debardeleben during the interview.

We immediately started talking to the Harrier operating above us. Through the pilot's conversation, I could tell things were getting worse on the ground. We made the judgment to accelerate the mission, moving toward top speed as the pilot relocated.

The pilot on the ground indicated that "they're still going at us, and things are getting worse." He was clearly on the move. We had the grid of the original crash site, then got a new grid and realized it was much further away—he'd been moving the whole time.

If I had been flying a SEA KNIGHT, by the time I got the new grid information and flew for 40 minutes under those conditions, I would have been relatively exhausted from holding the controls and getting shaken the whole time. On the Osprey, I'm on autopilot. I can take a sip of water, assess everything, and listen clearly to what's happening.

The V-22 is very quiet in airplane mode so we can hear the radios well, whereas the SEA KNIGHT's noise would make it difficult. The grunts in the back could look at a moving map display for situational awareness as we approached the coastline. They stayed relaxed and comfortable instead of being shaken around, which usually makes you groggy.

We descended from 500 feet to 200-300 feet about ten miles off the coast to stay below radar. Looking at the coastline, I expected Libya to look like Djibouti with a dark profile, but it was lit up with the electricity grid. So we picked a dark spot.

Another advantage over the SEA KNIGHT was the actual navigation system. Previously, I'd been holding a map with a flashlight trying to figure things out while someone else was flying. With terrain guidance, you can rapidly assess the terrain and plan your route. In the V-22, you get all the data right in front of you—it's like having a smartphone versus a dial telephone.

They gave me a new grid, and I looked for landmarks, a large town full of lights that was probably the area the downed pilot was running from. I adjusted course left, went through the dark area, and marked waypoints on my map.

The Harrier built a picture for me, describing what I would see, what the road looked like, and where he was positioned. The F-15 and F-16 were back on station, talking to the downed pilot. They knew him personally and encouraged him to drink water and stay calm.

Earlier, there had been commotion, people and cars chasing him. The Harrier used various means to scare people away. That's when he said "say goodbye to my wife." I could hear in his voice that things were getting serious.

Having people overhead in a calmer environment was invaluable. They asked if we had direction finding capability, then cued us as the needle swung over my navigation display, confirming he was in the area we were headed toward.

The joint quality of the operation was crucial. We're joint enough in terminology and techniques that we can interoperate effectively in these situations. I learned later that the downed pilot and I had attended introductory aviation school together in Pensacola eleven years ago—you never know how small the world will get.

One of the Osprey's best features is that it's quiet on the outside in airplane mode. Nobody hears us until we've passed them. It sounds like a whisper rush. At night, you won't detect anything. The SEA KNIGHT or any helicopter can be heard from 10-15 miles away.

As I approached, the pilot was very quiet, but I could hear dogs barking in the background, helping me envision his location. They reported vehicles pushing northwest, but nobody knew his exact position and things were looking good.

The landing zone wasn't well described, just "land by this road." We have an inertial navigation system that plots our position precisely in time and space. When we set up to land, we get a velocity vector showing our movement without having to look outside.

This is an awesome capability. In traditional helicopters, looking at dust and airflow can be confusing. It looks like you're moving backward. Landing in the desert has always been one of our hardest tasks. The Osprey has fixed this. We can land manually using the system, or the aircraft can fly itself to hover and land directly down. Both methods work 100 percent of the time.

As we transitioned to helicopter mode and began landing, the noise came on—and it's loud. Immediately, the downed pilot started yelling on the radio: "Don't leave me, I hear you." I responded, "We got it, we know where you are, we're coming. Send your flare up."

It was a bright night with few clouds, not ideal conditions for this type of mission. As he started talking, my crew chief finally said, "I see him." We got a sparkle from the F-16 marking the spot.

We landed about 20 feet from him. He jumped up, hands up, with no radio or pistol visible. He said he never even pulled his pistol from his holster. He ran to the aircraft, jumped on, sat down, fastened his seatbelt, and said, "I'm ready to go."

The grunts spread out to secure the zone momentarily. The crew chief ran out, grabbed him, and told everyone to get back on the aircraft. As the second Osprey, I didn't land but came up beside the first one. They picked up, and together we departed.

The mission's success was due in part to significant training. We trained for seven months as a team for this type of operation. Every time we had a chance whether on the boat, in Djibouti, wherever we were—we practiced TRAP missions. It's the one way we can coordinate the grunts in back with the jets overhead and successfully execute a landing in the zone.

SEVEN

2012 Visits: Focusing on the Return to the Sea

My visits in 2012 focused on the planning, execution, and lessons learned from Bold Alligator 2012 (BA-12), a major naval exercise focused on revitalizing amphibious warfare capabilities for the U.S. Navy and Marine Corps team.

Bold Alligator 2012 represented a significant milestone in redefining amphibious warfare for the 21st century. Rather than simply "storming beaches," it demonstrated how sea-based forces could project power throughout an extended battlespace, leveraging new platforms and concepts to create a more agile, responsive force.

As one officer noted, the exercise helped shape "a template for more strike capability" that would be further enhanced by future platforms like the F-35B.

Bold Alligator 2012: Testing New Strike Capabilities

January 26, 2012

By Robbin Laird and Ed Timperlake

During interviews conducted in Norfolk, VA and Cherry Point, we discussed with various combat elements involved in the Bold

Alligator 2012 exercise, focusing on preparations for the strike force off of the USS Kearsarge operating 16 Harriers.

A key interview was conducted with Colonel Andrew "Shorty" Shorter, Commanding Officer, Marine Aircraft Group 14; Lieutenant Colonel "Uber" Williams, OIC Marine Aviation Training System Site; and Lieutenant Colonel Shawn Hermley, Commanding Officer, Marine Attack Squadron 231.

Col Shorter During the interview.

Col Shorter explained the paradigm shift in using Harriers off large deck amphibious ships:

The use of the Harriers in this exercise is completely different from what we normally do. Normally, a Harrier squadron breaks down as a squadron (minus) and detachment that chops out to the MEU/ACE as six planes and nine pilots. The squadron (minus) remains behind to continue training pilots and as a deployable unit.

What's different in Bold Alligator is that we're taking two squadron (minus) that already have their Dets out dedicated to MEUs and deploying them on a large deck amphib to work as a single integrated strike force. Instead of the normal six-Harrier Det, the strike force will nominally have 16 Harriers from the two squadrons (minus).

We're testing something greater than two MEUs trying to aggregate assets into a single support element. We're experimenting with a much larger sea base support structure led by the amphibious fleet rather than the strike carrier. We're able to take these two entities that normally don't deploy on the sea base, put them on the sea base to be agile in reaction to requirements.

LtCol Williams discussed his role as Col Shorter's deputy and the exercise's operational scope: *The MEUs we put out every year have a somewhat small sampling of AV-8Bs, six Harriers, nine pilots, 74 Marines. That capability is limited compared to what we're putting out for Bold Alligator, which is an entire deck dedicated to Harrier operations with 16 airplanes.*

The first couple of days, we'll be shaping with integration with F-18s from Beaufort and prepping battlefields. Then we transition into a sortie generation phase demonstrating the flexibility and responsiveness that STOVL brings. If we're close to the beach, we can support that maneuver force going ashore, get back to refuel and re-arm, and get right back out.

LtCol Williams during the interview

We'll do long-range strikes partnering with shore-based F-18s, and simultaneously partner with our V-22s coming off the Wasp to do a long-range raid all the way up to Virginia. The flexibility demonstrated by having 16 AV-8s on a carrier dedicated to them means we're sweeping the breadth of the East Coast and the breadth of missions almost daily.

We're going back to the vision of a VSTOL force with squadrons of STOVL fixed-wing aircraft afloat and squadrons of V-22s afloat. The prox-

imity to the battle space and demonstrated capability in terms of speed, responsiveness, range, depth, and capabilities is phenomenal.

LtCol Hermley detailed the exercise's key challenges: *The big challenge was putting an entire MEB afloat and executing a landing while operating two Harrier squadrons on the Kearsarge alongside two other large deck amphibs plus CVs. We had to balance distance from shore that supports both air strikes with 16 Harriers and landing forces inserted by surface means.*

LtCol Hermley during the iInterview.

The close operational quarters of large deck amphibs created real challenges. Each ship's tower airspace is a five-mile bubble. When you put ships into close quarters, it's challenging to de-conflict air traffic as you close toward shore. We stopped flight ops several times because we needed to sort out overlapping tower airspace.

Flight deck manning limited us to a ten-hour flight window. Managing the flow of ship ops and air ops made that constraint problematic. We couldn't execute the sortie generation rate needed to maximize offensive air support effectiveness.

LtCol Williams emphasized the doctrinal implications: *During ESG-MEB employment we had multiple LHA/LHD ships and multiple amphibious ships with aircraft and surface craft. Each brought its own command and control, creating significant challenges in understanding how elements fit together and establishing clear chains of command.*

The Marine Corps and Navy have an opportunity during ESG-MEB size exercises to update our doctrinal approach to command and control and capitalize on emerging capabilities. We need to learn how to build ship positioning, phase offloads, and phase aviation operations.

LtCol Hermley concluded with a critical insight: *If we insert the F-35B into the current situation WITHOUT changes in mindset on both the blue and green side, it will run into the same problems. You need to adopt some techniques and procedures from the big deck carrier mindset to the evolving situation for large deck amphibs.*

This exercise helped us realize that we can't just do business like we normally do on a MEU and make it bigger. The ESG-MEB is not just an ARG-MEU on steroids. It's much more capable, and we can be smart about leveraging the increased numbers and qualities of assets to gain full capabilities.

The exercise demonstrated the potential for expanded capabilities. As LtCol Williams noted regarding the V-22 raid on Fort Pickett: *The large number of ships afloat and different capabilities in a broad sea base mindset dramatically expand the toolset. A raid may be originally driven by special operations forces launched from the sea base, but may quickly transition to a conventional mission from the same sea base.*

The flexibility of sea basing allows us to put the right force ashore with logistics and aviation support for the right duration. This provides broad breadth of influence over the battle space, giving commanders great flexibility and sustainability throughout the spectrum of operations.

Bold Alligator 2012 proved that the future of amphibious operations requires fundamentally new approaches to command and control, asset integration, and operational thinking and to set the stage for the eventual integration of F-35B capabilities into this evolving framework.

Changing the Mindset: "The ESG-MEB is Not an ARG-MEU on Steroids!"

March 21, 2012

By Robbin Laird and Ed Timperlake

We sat down with one of the two Harrier squadron commanders aboard the USS Kearsarge during the Bold Alligator 2012 exercise to discuss lessons learned from this major amphibious operation.

Question: Could you discuss the challenges associated with the exercise?

LtCol Hermley: *The big challenge was putting an entire MEB afloat and executing a landing. We had to operate two Harrier squadrons on the Kearsarge alongside two other large deck amphibs plus CVs, all within the same operational air and sea space.*

Operating two squadrons while getting the ship's company and deck crew to work seamlessly was difficult. We also had many young aircrew and enlisted Marines. When I asked who had never operated aboard a ship before, almost every hand went up.

Question: What other challenges emerged in executing BA-12?

LtCol Hermley: *A key challenge was balancing distance from shore to support both air strikes with 16 Harriers and landing forces from the ship's well deck. For surface operations, the ship was closer to shore than ideal for air strikes.*

The close operational quarters of large deck amphibs created major problems. Each ship had to maneuver in very close quarters within their sea space to stage near the beach for surface movement.

The tower's airspace control for large deck amphibs is a five-mile bubble. When ships operate in close quarters while closing toward shore, de-conflicting air traffic becomes extremely challenging.

Weather also complicated operations. Harriers need strong wind for heavy takeoffs but less wind for landing, requiring constant ship maneuvering. Because ships were positioned closely together, we stopped flight ops several times to sort out overlapping tower airspace.

The previous Bold Alligator was simulated and it is important to realize that until you do live ops, you don't see these problems or develop real-time solutions.

Comment: The exercise highlighted the need for better C2 and Air Traffic Control capabilities when shifting from ARG-MEU to ESG-MEB operations.

LtCol Hermley: *Absolutely. We need improved proficiencies for air traffic control and more time to work through ATC challenges. The close proximity of large deck amphibs, airspace control proficiency levels, and radar degradation issues all contributed to the problem. We need better systems and backup capabilities for ESG-MEB sized operations.*

Flight deck manning is another issue. We're limited to a ten-hour flight window, and managing competing priorities within that constraint, especially balancing daytime movement operations with nighttime strike missions, prevented us from achieving optimal sortie generation rates.

We didn't really integrate with the large deck carrier 70-80 miles away, though working with land-based Hornets from Beaufort went extremely well.

Comment: There's a significant challenge moving from ARG-MEU mindset to ESG-MEB approach.

LtCol Hermley: *Exactly. If we insert the F-35B into the current situation without changing mindset on both Navy and Marine sides, it will face the same problems we had operating fixed-wing jets off large deck amphibs. We need to adopt big deck carrier techniques and procedures for evolving large deck amphib situations.*

Question: How do you see the way ahead?

LtCol Hermley: *Most experience lies with ARG-MEU operations. Breaking that cultural mindset will be crucial. Three big deck amphibs bring different capabilities to the fight.*

This exercise showed we can't just do normal MEU business and make it bigger. The challenges we encountered, airspace control, sea space control, deck battle rhythm, flight windows, setting priorities, proved that ESG-MEB is not just ARG-MEU on steroids. It's much more capable, and we must be smart about leveraging increased numbers and asset quality to gain full capabilities.

The Impact of the Osprey on the Expeditionary Strike Group: "There is a Tsunami of Change Coming"

April 2, 2012

By Robbin Laird and Ed Timperlake

Following the Bold Alligator 2012 exercise, we interviewed LtCol Boniface, the Osprey squadron commander, to discuss how the MV-22 Osprey fundamentally redefined Expeditionary Strike Group-Marine Expeditionary Brigade (ESG-MEB) operations.

Two stark images illustrated the transformation since the last major amphibious exercise in 1996. In 1996, airborne forces conducted raiding functions from traditional platforms. In 2012, a raiding force led by Ospreys launched from a supply ship, conducting an assault raid 180 miles inland from the sea base with allied forces observing.

As Dutch naval officer Lt. Commander Pastoor noted: *We had Dutch observers and they were very impressed with the game changing capabilities of the Osprey in terms of range and speed. Normally, in such an exercise we would take the beach and operate 30 miles inland. With this new capability we can operate throughout the entire battle space and move forces as if across a chessboard.*

The second transformative image was the Osprey landing on a T-AKE supply ship before conducting the raid. This capability allows aviation assets to connect supply ships with combat ships, enabling more efficient use of combat vessels. Supplies can be reconfigured between ships while the MV-22 provides rapid delivery capabilities, enhancing the battle group's speed and agility.

Question: How important is the V-22 in defining the operations of the ESG-MEB in Bold Alligator 2012?

LtCol Boniface: *The comparison of the MV-22 to the CH-46 is a dead argument. The MV-22's sheer capabilities are causing us to rethink how we perform expeditionary operations from the sea.*

Traditionally our MEU concept focuses on a radius of about 100 nautical miles. With the Osprey's speed and range, why can't we change this to 500, 1,000, or even 1,500 miles?

I think of it like chess. A traditional ARG-MEU is like moving a pawn one space at a time. We now have the ability to move a knight, bishop, or rook and attack 180 degrees toward the enemy's rear. We can go directly after the king. The speed, range, and reliability of the MV-22 allows this.

There is a tsunami of change coming when we talk about the ability to fight an enemy and support Marines ashore. We can increase our area of operations exponentially because we can spread out our ships with an aviation connector that moves Marines tremendous distances in very short time.

The Osprey enables leveraging other aviation assets, AV-8B Harriers, CH-53s, AH-1Ws, and UH-1Ys, to support the Marine Air-Ground Task Force (MAGTF). Instead of moving 50-100 miles ashore, forces can now operate 200-500 miles inland with increased speed, range, and lethality.

Fire support concerns are addressed through Harriers and the future F-35 concept, along with Carrier Strike Group integration, providing coverage until organic assets like artillery can be positioned.

The T-AKE supply ship represents more than just logistics support. It can function as an enhanced Marine Aviation intermediate maintenance department afloat for an ESG. From an aviation standpoint, the ability to repair, reissue, and supply major components will be essential to ESG success or failure.

For the MV-22, this capability should expand to engine repair/overhaul, prop-rotor gearbox maintenance, conversion actuator repair, and having necessary technicians aboard to support sustained littoral operations without reaching back to the continental United States.

As Boniface underscored: *With legacy aviation assets we had to think inside the ARG-MEU 100-nautical-mile operational box. We have to get out of this mindset,* Boniface emphasized. *We're starting to operate a more disaggregated ARG-MEU, relying on the MV-22 as an aviation connector. Now we can move from our operational sphere of a few hundred miles to more than 1,000-1,500 miles in our area of influence.*

The Osprey's increased capability means it's handling more Passengers, Mail, and Cargo (PMC) requirements among ARG shipping. While the MV-22 can support both the MAGTF and Navy in bad weather over long distances, this creates new challenges for naval aviation connectors, as the SH-60 struggles to keep up with these extended distances.

As distances increase, getting sailors and Marines to appropriate medical care within the "golden hour" becomes crucial. The MV-22 can accomplish this; the SH-60 cannot over long distances. From a search and rescue standpoint, the MV-22 becomes the only platform capable of rescuing downed pilots over extended ranges.

Boniface underscored the follow-on effects from the MV-22 deployments: *How do we command and control this monster?* Boniface questioned. *Some LHD-class ships aren't ready. They don't have the band-*

width and throughput for command and control when you've got assets 1,500 miles away. Our command control needs to be bulletproof, and it isn't right now. The Aviation Combat Element comes with big intranet requirements, and 100-megabit LHD networks will struggle to keep up, especially as the F-35 comes into play.

Bold Alligator 2012 featured every East Coast amphibious ship within 150-200 miles of each other — all four large deck amphibs, support ships including the Lewis and Clark, the new LPD-17, and allied ships like the French Mistral.

The exercise proved the Osprey's remarkable capabilities: operating from off Jacksonville, North Carolina, the aircraft could reach Washington, D.C., in 45 minutes and return on a single fuel load with substantial cargo and personnel, without touching a forward operating base or tanker, while maintaining safety margins.

The successful T-AKE landing was conducted by mid-grade captains rather than senior officers, proving the concept's practical viability for real-world operations "at night and in bad weather." The USNS Robert E. Peary provided 5,000 pounds of fuel while the Osprey maintained a 10-percent torque margin and lifted 12,500-13,000 pounds total, including 7,000-8,000 pounds of cargo.

For Boniface, the Osprey represents a paradigm shift from traditional amphibious operations. Rather than simply assaulting beaches, forces can now maneuver within and over the entire battlespace, inserting, moving, and withdrawing forces with unprecedented flexibility. The Lewis and Clark-class ships offer tremendous untapped capability, especially in supply and repair functions.

As LtCol Boniface concluded: *It isn't a wave of change; it's a tsunami of change coming.* The MV-22 Osprey has fundamentally altered how expeditionary forces operate, demanding new operational concepts, command structures, and support systems to fully realize its revolutionary potential.

The Evolution of the Osprey: "We are No Longer a Bar Act"

September 9, 2012

The Osprey has been deployed for five years in combat. Starting in September 2007, the Osprey was deployed to Iraq and then later to Afghanistan. It has several years of at sea experience as well.

The operators – pilots and maintainers – have been part of the "testing" of the aircraft as it has developed under combat conditions. With new platforms, one reaches a point that the academic testing needs to stop and the real world operational use needs to take the project forward. This was clearly the case with the Osprey.

In a discussion with LtCol Brian McAvoy, the Commanding officer of VMM-264, a good sense of the progress of the Osprey under evolving operational conditions was provided. McAvoy was a CH-46 pilot for nine years and then transitioned to the V-22 in 2004. He went on the third Osprey deployment to Iraq in 2009 and then to Afghanistan for the third deployment there in 2011.

LtCol McAvoy provided insight into the differences in how the Marine Corps was able to approach the Osprey and its use in the two deployments as well as to provide insight from flights to participate in the Farnbough Air Show in 2006 and then again this year. In a fundamental way, McAvoy and his experience sheds light on the evolution of the Osprey as a weapon system used by the USMC in operations, both non-combat and combat.

Question: What were some of the major differences between the flights made in 2006 and 2012 with the Osprey to participate in the Farnbough Air Show?

LtCol McAvoy *In 2006, the flight to Farnbourgh was the first transatlantic flight of the Osprey. We flew 2 Ospreys to Farnbourgh and it was challenging. One of the aircraft made it all the way to the U.K. without an issue. The other V-22 experienced "an engine compressor stall" and was forced to make a precautionary landing at a U.S. military base in Iceland. We accomplished the trans-Atlantic mission, but much of what we did was "a first," so we were learning along the way."*

Fast forward to 2012, with multiple deployments under our belts and over a hundred thousand hours behind the Osprey Community, the transatlantic flight was a totally different experience.

We were required to have 4 Ospreys in Farnbourgh, so we started out with 6 flying from New River to St. John's Newfoundland, where we stepped off. The aircraft were all configured with three internal fuel tanks to give us about six hours of fuel. This has become standard operating procedure for long-range flights.

In St John's, we were joined by three KC-130Js; one tanker, one trail maintenance and one back-up. The next morning all aircraft took off without an issue. One hour into the flight, the back-up aircraft (2 MV-22s and 1 KC-130J) returned to St John's.

The rest of the package took approximately 5,000 lbs of fuel and continued on the 5.5 hour leg to Lajes, Azores. We stayed overnight in Lajes, and the next morning we again departed without issue. Four Ospreys and two KC-130Js flew directly to Farnbourgh from Lajes, Azores in about 6 hours. We had no significant maintenance issues along the way.

LtCol McAvoy after the interview at New River.

Question: How would describe the difference from 2006 to 2012 at RIAT and Farnbourgh?

LtCol McAvoy *In 2006, it felt like we were a bar act. It was challenging to get there and we were seen as oddities. In 2012, we were flying a plane with years of combat experience. We were no longer were a bar act, but war fighters flying and maintaining a key combat capability.*

We had a story to tell. We had folks with hundreds of hours of operational experience behind us. It was a proud moment for me as the CO. I was the senior guy and I had no majors; it was all captains, all young guys who really shined in explaining the aircraft to others.

In 2012, it was the young, but very experienced enlisted Marines and officers who told the MV-22 story, and their stories were based on actual combat.

Question: Let us focus now on the differences from the 3rd deployment to Iraq and the 3rd deployment to Afghanistan, a gap of only two years. How would you describe the differences?

LtCol McAvoy: *We were focusing on the basics in Iraq. In my squadron at the time, no one had deployed in combat with this aircraft. Some had deployed with other type model series, but this was the first with the V-22.*

In effect, Iraq formed a foundation for follow-on operations and we just gained experience from that deployment and minimized risks in later deployments. We grew together as a team and built a solid foundation during that Iraq deployment. When we went to Afghanistan last year, we leveraged that foundation.

We were familiar with the aircraft in tough operational conditions. We were able to expand the box exponentially. We moved on from logistic support to doing Named Operations. We have become a high performance assault platform. We are doing large movements day and night into the worst zone, hot zones, places where we know that we are going to have contact with the enemy.

In other words, as the experience grows from the team, we can expand what we can do with the aircraft. And the young guys with just V-22 experience are pushing the envelope.

During our time in Iraq, we used largely helicopter tactics. In Afghanistan, we were using V-22 tactics. The military as a whole is now willing to exploit the V-22 capabilities and not just treat it like it was a helicopter.

EIGHT

2013 Visit: Ospreys, Harriers and the Future

The visit in 2013 highlighted a Marine Corps aviation component in transition, developing new capabilities and tactics to support distributed operations across challenging environments while leveraging technological advances to enhance operational effectiveness.

An Update from VMX-22: USMC Aviation Works the Future

July 12, 2013

In this interview, Col Michael Orr, the CO of VMX-22. discussed the upcoming transfer of the squadron from New River, North Carolina to the Yuma Marine Corps Air Station in 2015. Col Orr provided updates on the squadron's activities, including testing the Osprey for Carrier Onboard Delivery (COD) and developing UAS capabilities for the USMC.

Col Orr: *We've talked about the importance of moving to Yuma and why it's crucial for Marine Aviation that this squadron collocates with MAWTS-1. We understand the importance of developing new systems alongside new tactics and how such integration will accelerate positive change for Marine Aviation. Yuma is the right place to thoroughly test our new aircraft and weapons.*

We're laying concrete for the facility, expected to be complete in 2015. However, we're not waiting for new facilities before starting the transformation. This summer, we'll establish our first detachment in Yuma, beginning with unmanned aircraft systems operators and Marine Aviation Command and Control System Marines.

VMX-22 hasn't done operational testing on these systems before. The Marines we're sending are hand-picked entrepreneurs and innovators who will develop our roadmap for future integrated testing. They'll conduct operational testing and limited objective experiments to test new payloads and platforms.

Question: There's a clear shift from UAS use in Iraq and Afghanistan toward a new approach. What does this mean for Marine Corps Aviation's future?

Col Orr: *The operating conditions in Iraq and Afghanistan were anomalies compared to how we expect the future Marine ACE (Aviation Combat Element) to support the MAGTF (Marine Air-Ground Task Force). Operating from fixed forward-operating bases isn't how we see tomorrow's wars unfolding.*

We think about how the ACE will support widely distributed operations, particularly in the Pacific. How do we support small forces dispersed over wide areas from both sea-based and shore-based facilities?

Our current unmanned systems are primarily line-of-sight platforms emphasizing traditional Intelligence, Surveillance, and Reconnaissance missions. Marine planners have identified critical capability gaps, such as conducting cyber and electronic warfare, which require beyond line-of-sight capability.

The team we're sending to Yuma will focus on addressing capability gaps, developing proper equipment, and working with MAWTS to develop tactics supporting future MAGTF operations. The continuing maturity of the Osprey is also key to this evolution.

Col Orr: *We're clearly concerned with the shortfall in amphibious ships. Both the CNO and Commandant are focusing on better leveraging all parts of the seabase. The question is how we can best use our sea-based mobility to support joint operations, including logistic ships and joint high-speed vessels beyond traditional amphibious shipping.*

I see the Osprey as a key enabler linking all nodes of our future sea-based power projection capability. The Osprey provides the range and flexibility to translate this sea-based vision into reality.

Question: Could you update us on your work with the

USN evaluating the Osprey as the Carrier Onboard Delivery platform?

Col Orr: *Assisting the Navy with the COD military utility assessment has been a significant effort involving most of the V-22 flying portion of the squadron. We've conducted four at-sea periods on both the USS Bush and USS Truman to assess the Osprey's viability as a COD platform.*

We put this event on the cover of our 2013 book on Pacific operations. Col Orr was the pilot of the Osprey seen on the cover of our book.

For our last evolution, we deployed with a VRC logistics squadron to Mayport, Florida to replicate the traditional COD role as closely as possible. The assessment's purpose was determining the Osprey's impact on aircraft carrier cyclic air operations.

We know the aircraft can perform well moving people and parts to ships. It was designed for that. The question is what impact Osprey operations will have on cyclic air operations, which are a finely orchestrated aerial ballet. From my perspective, the assessment went very well. We completed all necessary evaluations, and the crews and equipment worked magnificently.

Question: Could you elaborate on your thinking about UAS evolution for the USMC?

Col Orr: *LtGen Schmidle asked me to head this year's unmanned aviation system Operational Advisory Group (OAG). The OAG allows fleet operators to provide input to requirements teams who direct funding for current and future systems.*

Unmanned aviation is ripe to take advantage of recent technological developments. We've developed recommendations to improve our unmanned community in terms of equipment, training, and operator recruitment.

As we retire the EA-6B Prowler, how does the Marine Corps leverage the electronic warfare expertise in that community? Unmanned systems represent the next frontier for rapid innovation in cyber and electronic warfare. The MAGTF Electronic Warfare concept combines existing signals intelligence from ground-based sensors with cyber and electronic warfare techniques. Unmanned systems will play a key role. They're persistent and have the range, size, weight and payload for various tasks.

In the unmanned world, survivability takes on different context. Perhaps our unmanned systems' purpose isn't avoiding detection, but causing enemy reactions, like activating air defense systems. We can leverage unmanned systems for cyber and electronic warfare techniques.

Long-term, we're committed to joint solutions developing future platforms serving various joint requirements, including unique MAGTF needs. Mid-term, there may be opportunities to utilize excess UAS capacities as we draw down from Iraq and Afghanistan. You can use almost any airborne platform for cyber and EW, but unmanned systems provide advantages through persistence and ability to cover wide geographic areas where we wouldn't place manned systems.

Future aviation units might receive missions like air-to-air or air-to-surface warfare and use mixed manned and unmanned platforms. It's not hard to imagine an F-35 operator providing tactical command of both manned and unmanned systems supported by an entire ground team. While not technically feasible now, there's no reason it won't be achievable soon.

I'm excited about this unit's role in shaping Marine aviation's future. We're bringing together the right mix of people, equipment, and ideas to transform how we conduct operations. Marines have always been innovators, and I see this orga-

nization poised to dramatically improve how Marine aviation supports the MAGTF and Joint force.

The Osprey and Innovation: Breaking the Mold

August 5, 2013

During a visit to New River, Major Frank "Robo" Rhobotham, VMM-364 Remain Behind Element Officer in Charge, highlighted how the Osprey has forced significant culture change in the USMC and shaped new combat approaches.

Rhobotham discussed the remarkable speed of the Special Purpose MAGTF concept. It went from inception to deployment in just eight months. The force demonstrates exceptional flexibility with its light footprint, as he underscored: *With a six-ship Osprey force supported by three C-130s we can move it as needed. The three C-130s are carrying all the support equipment to operate the force as well.*

This flexibility offers Combatant Commanders and U.S. defense officials a major strategic and tactical tool for rapid support and force insertion in today's global environment.

The deployment planning process required answering critical questions about mission parameters. As Major Rhobotham explained, the team had to consider whether they were "going austere" or working from prepared zones, flying 100 or 500 hours monthly, and operating environments that affect aircraft performance and endurance.

He compared it to automotive use: *If you buy a truck from the Ford dealership, and you drive it around LA, you're going to get 150,000 miles out of it. You take that same truck and you attempt to run the Baja 500, you won't make it past the first day.*

Similarly, operating from paved runways in airplane mode extends aircraft life compared to helicopter operations in harsh, dusty environments.

When Africom expressed interest in a six-plane V-22 force, the planning focused on Africa's diverse environments. Though Mali dominated news during planning, the team examined western and northern Africa, with Libya fresh in memory.

The Osprey's refueling probe capability transforms operational reach. As Rhobotham underscored: *We're no longer limited to how far the ship is willing to steam in one day. Now we're limited to how much can that tanker hold?*

This enables self-deployment across oceans. The SP-MAGTF flew across the Atlantic in April rather than being airlifted by the Air Force. The force's mobility provides significant operational flexibility. The Major underscored: *If political turmoil erupts in any country, it doesn't take much to completely pack up and move.*

Major Rhobotham clarified that the SP-MAGTF complements rather than replaces MEUs, comparing them to tools in a toolbox: *It is similar to having both a screwdriver and you've got a drill in your toolbox; that drill is a lot like the MEU. It's a lot more powerful, it can go a lot faster; it can do a little bit more powerful things. But it doesn't mean you need to throw away your screwdriver. The SP-MAGTF offers a lighter footprint for situations requiring government attention. Unlike traditional deployments requiring infrastructure establishment and diplomatic agreements, Now I can fly in the force; stay until I wish or need to depart.*

Beyond operational innovation, the Osprey has catalyzed cultural change through a new generation of maintainers. These young Marines possess remarkable information processing abilities that enable creative problem-solving.

Major Rhobotham highlighted: *The new generation grew up with such an influx of information that they are able to process information in ways that are a challenge for me, and for my parent's generation are impossible. And it makes them amazing mechanics. Unlike previous generations who relied on encyclopedias and limited references, today's Marines are very used to opening up a source and saying well, I can't prove that this information that was published by so-and-so on this website's true. And they'll grab something 180 out, cross-reference it and make an assessment.*

This capability proves crucial for troubleshooting complex systems that don't always fail consistently. These mechanics creatively combine information from multiple publications, developing new procedures that get incorporated into official processes. Major Rhobotham added: *I attribute it to the way their brains and the way they're socially trained even from a young age to look at information and not necessarily believe that just because it's written in a book it's the end all, be all.*

In short, the Osprey represents more than technological advancement. It embodies a fundamental shift in Marine Corps operational thinking and maintenance culture. Through rapid deployment capabilities and the intellectual agility of a new generation of maintainers, the Osprey is reshaping how the Marines approach 21st century challenges. This thinking process proves crucial for operating aircraft as complex as the V-22, demonstrating how innovation extends beyond hardware to encompass human adaptation and cultural evolution.

New USMC Aviation Assets, Increased Maintainability and Combat Capability at Sea

August 21, 2013

The U.S. Marine Corps' upgraded Huey and Cobra helicopters represent more than simple aviation improvements. They fundamentally reshape the force's operational capabilities, particularly in maritime and expeditionary environments.

Major Nathan Hill, Executive Officer of Marine Light Attack Helicopter Squadron 269 (HMLA 269) at New River Air Station, provided insights into how these upgrades transform combat effectiveness aboard amphibious vessels.

The most significant advancement lies in the upgraded helicopters' design philosophy: over 80% parts commonality between the Huey and Cobra platforms, including shared engines. This integration creates cascading benefits throughout the logistics chain, from maintenance operations to supply management, fundamentally altering how Marine Expeditionary Units (MEUs) operate at sea.

Major Hill's experience transitioning from the legacy UH-1N (November class) to the upgraded UH-1Y (Yankee class) illustrates the dramatic capability enhancement. Flying the November class required constant physical and mental strain, with aircraft regularly operating at maximum gross weight limits of 10,500 pounds. Every flight presented potential engine over-temperature risks, and single-engine failures meant immediate descent and landing due to physics constraints.

Major Hill after the interview.

The Yankee class transforms this operational reality. With a maximum gross weight of 18,500 pounds and typical operations at 16,000-17,000 pounds, the aircraft operates well within performance envelopes. The new engines provide substantial safety margins. During precision approaches that previously left no power reserves, pilots now maintain 20% power margins for emergency maneuvering.

Mission capability has expanded dramatically. Where the November class carried maximum two personnel at the cost of ordnance, the Yankee regularly carries six personnel without sacrificing weapons loads. The platform has returned to its assault support roots while maintaining enhanced ISR and weapons support capabilities.

The technological advancement is immediately apparent in the cockpit. Legacy aircraft required constant pilot vigilance through manual system monitoring which is a mentally exhausting scan pattern to detect malfunctions. The Yankee class features four multi-function displays (MFDs) powered by dual mission computers that actively monitor all systems and alert crews to anomalies.

Yankee class helo at New River, July 2013.

This integration liberates substantial pilot cognitive capacity from aircraft management to mission focus, directly improving support quality for ground forces. Whether providing close air support or inserting Marine teams, pilots can concentrate on tactical requirements rather than aircraft limitations.

The engine commonality between upgraded Hueys and Cobras creates profound operational advantages aboard ships. Traditional MEU deployments required approximately 20 engines — 10 for each aircraft type — consuming valuable shipboard space and requiring specialized maintenance personnel for each engine variant.

The H-1 upgrade reduces this requirement to 10-12 engines total while eliminating the need for separate support equipment, specialized tools, and dual-qualified maintenance personnel. One maintainer can now service both aircraft types, effectively doubling personnel efficiency while reducing logistics footprint.

This flexibility extends to operational deployment patterns. MEU commanders gain unprecedented ability to distribute assets across multiple ships without losing maintenance capability. Cobras and Hueys can be cross-decked between vessels while maintaining full operational readiness, enabling disaggregated operations previously impossible under legacy constraints.

Amphibious deck operations present unique challenges distinct from large-deck carrier operations. Three critical factors govern operational tempo: ship's crew limitations, squadron flight crew availability, and regulatory flight time restrictions.

Flight deck crews operate on 12-hour shifts with limited personnel pools, while flying squadrons can sustain approximately 18 hours of continuous operations with H-1 upgrades. Weather delays compound these limitations because extended periods of flight operations suspension due to storms can degrade crew currency requirements, forcing additional qualification flights before combat operations can resume.

The Indonesia tsunami relief operations in 2005 exemplify this challenge. After 25 days of humanitarian assistance missions, the MEU's Harriers lost combat readiness due to currency lapses, requiring 2-3 days of intensive flight operations to restore combat capability. Future assets like the F-35B promise reduced currency requirements, maintaining combat readiness throughout extended humanitarian missions.

These upgrades represent a fundamental shift in amphibious warfare capability. Enhanced reliability, maintainability, and performance directly translate to increased combat power projection from maritime platforms. The ability to maintain disaggregated operations while preserving full maintenance capability aligns with modern distributed maritime operations concepts.

The H-1 upgrades demonstrate how seemingly technical improvements can create strategic advantages. By solving basic engineering challenges, such as parts commonality, system integration, and performance margins, the Marine Corps has enhanced its core mission: delivering combat capability from the sea.

In other words, the transition from legacy to upgraded systems illustrates broader military transformation principles. Success depends not merely on individual platform improvements but on how new capabilities integrate across the force structure. The H-1 upgrades' impact on MEU operations provides a model for evaluating future aviation modernization efforts.

Cobra at New River, July 2013.

Major Hill's insights reveal that the most significant military capabilities often emerge from addressing fundamental operational constraints. The upgraded helicopters don't just fly better — they enable entirely new operational concepts while dramatically improving the reliability and sustainability of traditional missions.

For MEU commanders, these upgrades translate directly to enhanced operational flexibility, reduced logistics burden, and most importantly, the ability to maintain combat readiness throughout extended deployments regardless of mission type or duration.

The 22nd MEU ACE Commander on Ospreys, Harriers and the Future

August 7, 2013

New USMC aviation assets represent more than simple upgrades. They fundamentally reshape force capabilities through improved maintainability and operational flexibility.

LtCol Schoofield, 22nd MEU ACE Commander, brings unique perspective as both experienced Harrier and Osprey pilot with over 1,000 hours in the AV-8 and combat experience in Iraq and Afghanistan.

He commented: *They're strikingly similar. They have more in common than you might think. Both are 'powered-lift' aircraft with ability to take off and land within their own dimensions then cruise on wing-borne flight. Both* require careful gross weight management through fuel jettisoning to maximize performance at landing points.

- Offensive Air Support (Harrier): *Even the most complex close air support missions are pretty scripted, rigid timing for takeoff, landing, ordnance, and procedures. A long Harrier day might be 5-6 hours flying.*
- Assault Support (Osprey): *Virtually the polar opposite — extreme flexibility and very dynamic missions. A long Osprey day is 12-14 hours strapped in. You're communicating plans with 7 other people across two aircraft, and plans always change.*
- Field conditions constantly shift: *You're expecting 12 people and get 18, or expecting 18 and get 12 plus dogs, interpreters, or extra equipment. Anyone flying assault-support knows this challenge intimately.*

Ospreys part of ACE for 22nd MEU seen at New River, July 2013.

I remember flying up the Potomac, literally passing a Black Hawk like it was standing still. They were doing 130-140 knots, we were doing closer to 300. The comparison highlights fundamental operational differences: Each Black Hawk has 3-4 crew carrying 12 soldiers slowly. We're moving 24 Marines with similar crew size at 300 knots ground speed. It's the difference between walking and driving. Combat operations are about putting force on target before the enemy moves or becomes more effective. The Osprey builds that advantage.

In short, these aviation upgrades represent a comprehensive transformation of USMC expeditionary capability. Through parts commonality, enhanced performance margins, and speed advantages, new assets enable more flexible, sustainable, and effective operations from sea-based platforms.

The combination of improved individual aircraft capability and simplified logistics creates multiplicative effects that fundamentally enhance the MEU commander's ability to deliver combat power across diverse operational scenarios.

Major Hutchings on Osprey Afghan Operations

July 31, 2013

By Robbin Laird and Ed Timperlake

In this discussion with Major Hutchings, we focused on the events involved in his Osprey coming under battlefield fire in performing its assault mission role.

Major Hutchings: *I've deployed multiple times to Afghanistan. The first time was in the summer of 2010, and that was a seven-month tour. We were the second V-22 squadron over there replacing the first Osprey squadron in theater. And then, I had a year back at home and then redeployed to Afghanistan again in January of 2012 for seven months.*

These were two completely different deployments. The joke was that the last deployment was not the GS deployment, the GS being the general support mission where all you do is go from FOB to FOB. Earlier, we were basically making logistics runs.

Comment: On this second deployment, your primary mission was an assault mission.

Major Hutchings: *Assault mission, full up. In our assault role, we are working closely with the other elements of USMC aviation. We worked closely with Hueys and Cobras, and on occasion with the British as well. The latter was a bit challenging because our doctrines and terminology is a bit different.*

The mission would be built around a CONOPS approach. And every CONOPS brief had a timeline slide. For multiple missions, we would do the assault with CH-53s and Ospreys. It could be say, a section of CH-53s or three Ospreys going to the same objective area with skids in the overhead providing the escort piece.

With regard to the timeline, the CH-53s or skids would take off first. We would take off up to an hour later, and be back an hour earlier than everyone else from the mission.

With the CH-53s, you can hear them coming for miles ahead. It is moving slower and lower. With the Osprey, no one can hear us or see us, and drop down in there right at the very end. That's the plain and distinct advantage we have over those guys.

Question: What are the main threats you faced when flying?

Major Hutchings: *The main threat in Afghanistan is small arms. So, you stay roughly out of small arms range. Because the Osprey maneuvers like a plane and flies at different speeds than the skids, it complicates the ability of the Taliban to target an entire assault package coming in. The training wheels were off of the Osprey, and by the end of my time during my second deployment, we were doing 9-10 named missions per week.*[1]

Question Could you discuss the events surrounding the Taliban attack on your Osprey, which damaged the aircraft, and then you had to fight your way back home? In other words, could you talk about the operation, which was recognized by the Distinguished Flying Cross?

Major Hutchings: *We had an a.m. crew and a p.m. crew. The a.m. crew was from 01 to 1300, and basically was split into pods and crew chiefs and half, and the afternoon crew started at 1300, went to 01. This provided a max amount of flexibility for the ground forces to do missions prior to sunrise, and after the sunset.*

All of the inserts and extracts will go prior to BNMT, which is before the sun comes up. Or after the sun goes below 12 degrees, below the horizon, so it's completely black.[2]

We planned for a two Osprey insert of ground troops to do a reconnaissance mission. You never know how hostile the territory will be when you engage. With the Taliban it is like the Whack-a-Mole game.

The landing areas were two sides of a river, and we were to do two inserts of Marines. The first insert came off without a hitch. It was different the second time in. Of course, we never land using the same landing pattern, but the territory was bit constrained so your options were limited for landing areas.

We were landing at night and as you land you are focusing on the terrain and navigating through the brownout challenges. You tend to be heavy, because they are always cramming the next guy or support gear.

Major Hutchings after the interview.

So what are the pucker factors coming in to land?

- *The first pucker factor is flying low over the mountains in the dark.*
- *The second pucker factor is the actual landing.*
- *The third is making sure we had a enough gas to get home. If not you have to plan for an alternative site for landing on return.*

We started taking fire from the ground. You could hear the bullets hitting the airplane, ting, ting, ting, ting, ting. By this point, we're committed to the landing because we're in the dust.

We continue to take rounds coming in. RPGs come through in the dust. The fuel lines are hit; the ramp will not open. The machine gun jams at the rear of the plane. The brakes don't work and other things cascade as well. The guys start jumping out of the back.

After the landing, the gauges tell us not to fly and the condition of the aircraft is such if that we were at an airfield we would have not flown the aircraft. Upon takeoff, we conducted controllability checks and the gearboxes are getting hot as well. We can't raise the gear because the hydraulic system blew out. We were flying in airplane mode with 40 miles to get to Camp Bastion.

The copilot (Dave Austin) suggested that we land on a ramp at Camp Bastion because our brakes were not working. They brought out a tug to bring the aircraft in and after we got out we could see the plane was riddled with ground fire.

We were safe but our thoughts were of Captain Haake's plane and concerned about his fate. Later we learned, that he too had taken ground fire and had to land at the FOB rather than making it back to Bastion.

With regard to the Osprey, our experience shows that it is a tough bird. And it's proven that it can take rounds of critical flight components and keep on flying. The redundancy built into the aircraft is crucial. The redundancy works. As long as the engines are spinning and the wheels are turning, it's going to make it somewhere.

Captain Haake on Osprey Afghan Operations

July 31, 2013

By Robbin Laird and Ed Timperlake

In this discussion with Captain Haake, we focused on the events involved in later receiving the Distinguished Flying Cross.

Question: We talked earlier to Major Hutchings about the events. You were the pilot of the second Osprey in the two-ship formation?

Captain Haake: *I was. Major Hutchings was the aircraft commander for the lead aircraft. I was the aircraft commander for the second aircraft.*

And when we came in for the second wave, that's when we had all the small arms, RPGs. Basically we came in from the South. It was anywhere from our 11:00 to 5:00. The only way I could describe it is it looked similar to a dance club in terms of illumination. So obviously it was a hot zone.

The call was made from the Cobra Huey Section in overhead above us to wave off, so I waved off, made a hard left turn, and reached back out to the holding area, which was about 10 miles away.

I assumed Major Hutchings was doing the same thing and didn't realize until a minute or two later that he landed and dropped his guys off and shortly after, he said he was heading back home as they had taken some rounds on the right side of the plane and they were reverting back to our home base, Camp Bastion.

Our agreement with the ground guys was to have no less than 34 Marines on deck and we were around half of the contingent so we had to land.

We worked with the Cobra Huey squad or Cobra Huey section to come in from a different direction. I didn't see any fires to north of the zone when we first came in. We decided to land just north of where we saw all the bad guys.

When we were about 150 feet, I heard Sergeant Moreland, who's my gunner on the back, with his 240 just laying down lead and so I knew right away that we were getting shot at.

We've been in situations like that a couple of times. You're pretty vulnerable at that altitude and that airspeed, so we just continued on down to the ground. We could see some of the fires off to our nose, which was then the original area where the fires were coming, but they were a non-factor.

I don't think they saw us, but we happened to pass right over the bad guys. Of course, they peppered the bottom and the right side of our plane, actually all sides, the bottom right left and rear end with multiple machine gun rounds and we landed on the deck.

The Marines in the back ran off. One of the crew chiefs said, "Hey, we got a casualty in the back. Looks like he's been hit in the leg." We hoped to have him back at Camp Bastion in less than 10 minutes and so we took off.

Immediately after I took off, I could feel that the controls weren't normal at all. The automatic flight control system in the plane had gone out because usually you can fly holding with two to three fingers. I was basically steering the plane at that point and I couldn't transition to airplane mode because after I started bringing the nacelles down, the airplane wanted to go down and to the right.

And then after about two miles out, we decided to stay in the helicopter mode. Then after about 2 miles from the zone, I noticed we had lost several thousand pounds of fuel and the thought was: We can't go back to Camp Bastion, which was 40 miles away. There's a FOB about 18 miles away, FOB Edinburgh.

I knew they had a medical staff at that FOB as well and so that's where we decided to divert to. Our biggest worry at this point was having fuel starvation, or running out of fuel.

Question: At this point, the race was on to get to the FOB prior to running out of fuel and you are flying in helicopter mode?

Captain Haake: *We were. Either the Cobra or the Huey, attached over the landing zone and then tried to join us at the zone because I thought we were going to either have an uncontrolled crash landing into the open desert or we're going to have to set it down because of fuel starvation.*

But the fuel leak slowed down after two of the main tanks were completely depleted. Two of the feed tanks that are in the nacelle engines still had fuel. With the projected fuel burn, we calculated that we would land with about between 700 and 800 pounds of fuel. We basically flew 50 to 100 feet of the deck in helicopter mode all the way back.

As I said, we could not transition to airplane mode. The plane wasn't able to go to airplane mode because of what turned out to be damage to what they call drive two which allows movement of the pitch on the prop rotors.

The right side wasn't getting any kind of movement because a round or bullet fragment had sheered the ridges on the drive two, which wasn't allowing a shift in the power. You have to have a lot more power as you transition to airplane mode; because of damage to the drive, I could not do that.

We did make it all the way back to FOB Edinburgh. It was a tough landing just because of controllability issues. We off-loaded the Afghan soldier who had been injured and got him medical help as well.

Question: After you landed, what did you assess to have damaged the Osprey, preventing it to fly in airplane mode?

Captain Haake: *We sustained damage to three systems. I not only had lost a hydraulic system, but I had physical damage to the actual prop rotor system. When you combine those two with damage to my automatic flight control system that meant that there three things preventing me from going to helicopter mode. Of course, we also had a major fuel leak as well.*

I can't think of any other aircraft in the inventory that I know of that can endure something like we experienced. We have significant redundancy built into the aircraft and a very well trained crew.

Nobody died that day, so we're definitely all lucky, and I think it was a combination of the aircraft and the entire crew that led to that result.

In addition to the two pilots, five Air Medals were awarded to the crew: Co-pilot Captains Austin and Vandenende; Crew chiefs Sgt Leist, Belleci, and Moreland; and LCpl Rhorer.

Maj. Gen. Robert F. Hedelund, 2nd Marine Aircraft Wing commanding general, presented Capt David W. Haake, Marine Medium Tiltrotor Squadron 365 MV-22B Osprey pilot, with the Distinguished Flying Crosses in the squadron's hangar aboard Marine Corps Air Station New River, June 28. (Photo by Cpl Manuel Estrade).

Sustaining the Osprey: Meeting the Challenge

August 15, 2013

The exit interview I did with Colonel Christopher 'Mongo' Seymour in the summer of 2013 during the week prior to his retirement, the hard hitting and well-respected Marine Corps leader provided a look back and a way ahead with regard to sustainment of the Osprey.

Question: A major challenge in fielding a new system is getting the supply chain up and working and getting the

inevitably maintenance problems sorted out. How have you worked through these problems?

Col Seymour: *There are three separate streams of activity which need to align to really get the new system up and running and integrated into operations.*

The first is getting the Marines committed to owning the system and learning how to fix "new" problems, which come up with a new system. The problems are different and have to be worked differently. You need to get the maintainers to change their culture.

Sorting out problems with the gearbox is a good example of what needed to be done. The gearbox on this airplane is very complex and central to its unique operational capabilities. The gearbox inside the nacelle turns a rotor, and they were chipping. This is high-end engineering. But it was chipping and when it did so maintainers put it aside and waited for a new part. This meant the fleet was going to be degraded.

The flight line needed to take ownership of the problem because a lot of it was self-inflicted gunshot wounds. Maintainers would look to blame someone else when they had a proper gearbox go bad. As it turns out, the, the technology required was to use isotropic oil that actually absorbs moisture out of the air, so if you have a gearbox that's not turning and boiling the oil out on a regular basis, it goes long term down. It's sucking in the moisture of the North Carolina Coast into the oil.

And the maintainers would leave it out on the flight line all opened up just breathing the air, and then when they finally got a part or piece, they try to fire it up and another gearbox would chip or another problem would manifest itself someplace else. It was an endless loop.

We took some ownership here on flight line, and shaped better maintenance practices, and to help industry. Once we got that proper gearbox moving back out of the red into the black, the internal culture of the community changed to become significantly more optimistic, you know. The maintenance man-hours required to change a proper gearbox initially was estimated at 1800 maintenance man-hours. We're doing it now in about 380. That's how good we got at it.

Col Seymour after the July 2013 interview.

Question: The first challenge then was simply to get the maintainers familiar with the plane, adjusting their practices and communicating better with industry to solve the gearbox problem. What were the other challenges?

Col Seymour: *The second thread is industry. The key challenge is to get industry working to solve the right problems, or the critical problems, which would make a difference.*

But as we focus on solving a core problem, there is the challenge of dealing with simpler problems, which emerge, which then slip between the cracks. Once industry is on a steady track, the PBL system incentivizes industry to drive down cost. They will not do this if they have little capacity to know what our operational costs for a system and its component parts are. We have to work with industry to get the reliability of key parts up and then the PBL system works quite well to shape an incentive system for industry to deliver parts on time and to improve costs.

The third thread is the government. In our case, the challenge is to get NAVAIR focused on solving the critical problems, and not just dealing with the low hanging fruit. When we got NAVAIR focused on solving the gearbox problem as the key, industry kicked in and worked the problem as a priority. The challenge is get those three strands to work together effectively – the USMC, industry and NAVAIR.

Seymour provided a number of insights into what was necessary to succeed in shaping a new culture for maintenance: *A key challenge is to get the maintainers to be stakeholders in the enterprise. For example, I had a gunny sergeant who said he preferred CH-46s to Ospreys. "He said that I'm a 46 guy." I looked him in the face and said, "Well, you're in the wrong unit, gunny. There are no 46 that's left here on the East Coast. You want me to get your orders to Okinawa, because, you know, that's the only place for flying 46s."*

I told him that you don't get a vote. This is not a democracy. Conway decided. McCorkle decided. Hagee decided. Amos decided, you know, we're going to fly the Osprey. It is what it is, so either embrace it or leave.

He then underscored where that Gunny is now in terms of working the Osprey. The same gunny now will brag about being, , the gunny who fixed problem X, Y, or Z in the maintenance department on a V-22. He owns it now. It's like okay, you know, it worked, it's normalized if you will, and that's why you see this, this growing success. Success begets success internal to the Marine Corps culture.

Seymour underscored the importance of industry and the SYSCOM listening to the Marines as they deployed the plane.

We had another problem with the gearbox where it attaches inside in nacelle with a component called the drag pin that, that holds that, that gearbox in place. There's a fitting around that drag pin that with vibrational wear and would fall out. And so you had to pull the whole gearbox out, replace this $25 bushing and put the whole gearbox back in at a cost of like 1800 maintenance man hours initially. So we did a lot of kicking and screaming at industry in and to the SYSCOM.

Are you guys, are you guys kidding us? You know, you want us to pull this $1 million component out to replace a $25 bushing at the expense of 1800 man hours probably equates to a million dollars just in labor cost, right? We don't pay labor cost obviously, but it's a good argument to make. Marine labor is free, but it comes on the backs of marines and it's demoralizing to the community.

Seymour also underscored the challenges of getting NAVAIR focused on solving critical problems, rather than what was easiest to solve.

NAVAIR has taken credit on briefing charts for dealing with low-hanging fruit. We have not fixed the gearbox but you got this. You know, what do you do? That's not going to do anything for me. That gives me, you know, a 0.001 percent bump in readiness. What's going to get me a double-digit bump in readiness is having the proper gearbox.

We finally got the SYSCOM focused on the gearbox, which, which actually focused industry on fixing the gearbox. Industry invested a lot of their own money, upfront money when they had to, with the multi-year PBL.

They could not get this gearbox into the PBL because industry saw it as too risky. The costs were all over the chart. They would replace one gearbox or repair one gearbox for $800,000 and repair another one for $50,000. And of course, NAVAIR and NAVSUP wanted a flatline PBL and industry understandably wouldn't take the risk. If you go for the low-hanging fruit, it might make you feel better for 30 seconds, but it's not going to solve the problem. Go after the hard, the big-bone items and that's what this gearbox was. It was a big-bone item.

He revisited the PBL issue when asked what is the key item, which needs to be remedied currently.

Blades, rotor blades, prop rotor blades. It drives me insane because in complexity, that proper gearbox is like splitting the atom and that blade sitting on top of it, it's like a 2 x 4. They focused all their energy on fixing the proper gearbox and they took their eye off the blades and, and now we got into this big bathtub on improper blades.

But it is absolutely solvable, and it's a PBL item. Industry is incentivized to improve reliability of the blade because every time they got to fix one, it costs them money. So they are going to work hard. You'll see probably in the next two or three years the blades won't be a problem anymore.

But we're going through this bathtub right now where they just weren't incentivized to focus on it. They were making more money dealing with the blades when they came in for repairs, but now that it is part of the PBL they will solve it.

NINE

2014 Visit: Reshaping USMC Expeditionary Capabilities

The 2014 interviews portrayed a Marine Corps in transition, leveraging aviation innovations to reshape its expeditionary capabilities while addressing the challenges of operating in more distributed environments across greater distances.

During the 2014 visits, I especially focused attention on the KC-130J and its core role for the USMC. It is a lifter, a tanker and with the Harvest Hawk variant, an early example of lift assets as weapons carriers, a trend which I expect to see accelerated in the years ahead in the airlift community.

The KC-130J in Afghanistan: In Support of U.S. and Allied Forces

July 28, 2014

During my visit to 2nd Marine Air Wing in June 2014, I had a chance to meet with several of the members of the VMGR-252 squadron.

In this interview, Major Mark Montgomery, a KC-130J pilot, discussed his experiences in Iraq and Afghanistan. He deployed in

Iraq from 2006-2008 and then in Afghanistan in 2012. One of the things that has always amazed me over the years in discussions at 2nd MAW is how matter of fact the Marines I have interviewed are.

You talk with them and in the course of the conversation accomplishments are mentioned not really highlighted as amazing achievements. It is a striking contrast to Inside the Beltway where one is constantly reminded of the importance of whomever you are talking with, notably at seminars.

Major Montgomery after the interview.

In this case, Major Montgomery highlighted one key difference between Iraq and Afghanistan. *In Iraq, we largely supported Marines; in Afghanistan we supported Marines, Aussies and Brits because our AORs were next to each other.*

With regard to the Aussies, the squadron provided various types of support such as transport of troops and Harvest Hawk Close Air Support.

With regard to the Brits, Major Hamilton mentioned an incident where they were tasked to provide Battlefield Illumination in support of a British Forward Operating Base: *We received the tasking, and the crew was airborne very rapidly with 100 flares on board and we operated over the objective area for 8 hours and was able to support what the JTAC wanted in that situation.*

Major Montgomery underscored that the close working relationship among the crew members is what made the flexibility and rapidity of re-tasking work: *We worked with the same group of guys and you learned to do things together rapidly.*

A difference in operating the Harvest Hawk configuration was highlighted by the pilot: *In our normal mission sets, we are taking troops or equipment to a certain point and dropping off or operating airborne to provide tanker support. With Harvest Hawk you are loitering to determine a strike position which needs to be VERY precise and in response to the demands from the JTAC.*

He also highlighted the importance of battlefield illumination (BI) and the various ways they operated in support of the forces on the ground with BI: *We did terrain denial, deception through faints, and direct support for insertion of helicopters and Ospreys.*

Evolving the Concept of Support with the KC-130J: An Interview with the Leadership of VMGR-252

July 30, 2014

The Osprey has been a dramatic addition to the USMC and has made it the only tiltrotor assault enabled force in the world. What can be lost in the picture is the role of the KC-130J.

The pairing of the two aircraft is changing how the USMC operates and thinks about the range and speed of operations. The pairing has allowed for the emergence of the Special Purpose MAGTFs, and to provide major support for Humanitarian Assistance and Disaster Relief (HA/DR) missions such as in the case of the recent Philippine relief mission.

The Ospreys being refueled by a KC-130J during the Philippine relief mission. The Marines twin the assets to provide for greater range and endurance in the mission. Credit Photo: LtCol Brown.

Earlier I had interviewed the CO of the Sumos – the KC-130J squadron which has just moved to Air Station Iwakuni and he indicated how the pairing worked together to re-shape USMC operations.[1]

The pairing between the KC-130Js and the Ospreys has brought an ability to shape organic modularity for long-range insertion of force in the region. We want to be able to provide for long-range vertical insertion throughout the region and to be able to deploy widely throughout the region as necessary. It is part of the operational dynamic and part of deterrence as well.

I would add that since the arrival of the Ospreys, about 2/3rds of our tanking requirement is to support the long-range assault support capability, which the Osprey provides. We can now power project vertical lift anywhere in the region for the MAGTF faster than before.

This is not just an abstract capability, but has been used with effect already in various situations, one of the most visible of which was coming to the aid of the Philippines as part of the shaping function to even set up a relief effort. And the ability to assist in the Philippines and to leave expeditiously (along with the USAF and USN engagement) was part of background to which the Philippine government re-opened its facilities to US forces in the region.

Put in other words, an agile military capability enabled a political bargain important to the Pacific strategy of defense in depth.[2]

As Lt. Col. "Sniper" Brown, the CO of the "Flying Tigers" or VMM-262 put it: *When the call to fly into the devastated areas came on Veteran's Day weekend, the challenge was to put together the ability to fly. We flew with the Sumos who kept us fueled and carried our logistical needs as well. But we needed to sort out where to go and what the priorities would be in the initial 72 hours."*[3]

But this capability did not happen overnight; innovation takes time.

It is easy to forget that the KC-130Js preceded the Ospreys by a few years, and it took time to sort out how to use the new capabilities inherent in the KC-130J and with these capabilities to evolve the concept of support.

The introduction of the new aircraft had its challenges. But the classic notion of lift and tanking associated with the C-130s broadened over time. The KC-130J not only has broadened its range of mission sets, but has provided a foundational capability for expeditionary support.

It is about supportable reach; not just reach.

When the crew could operate at night operations, the crew broadened its support to the Marines on the ground. When the Harvest Hawk was added, direct fire support to the Marines on the ground became part of the KC-130J operational capabilities, which in turn has established the notion that the airplane can become something different in the future as ISR and C2 engagement expands with the inclusion of UAVs and F-35 systems pushing data to the aircraft.

The point can be simply put: the KC-130J is becoming a significant part of a broadened concept to the Marine Corps assault team.

In an interview with the CO of VMGR-252 and the Executive Officer of the Squadron, the evolution of the role of the KC-130J was the focus of attention. The CO of the squadron is Lieutenant Colonel Scott M. Koltick and the Executive Officer is Major Ryan Pope, both are very animated, articulate and impassioned about what their squadron contributes to the USMC.

Lieutenant Colonel Scott M. Koltick and the Executive Officer is Major Ryan Pope.

The CO provided a succinct overview of how he saw the evolution of the KC-130J: *When we first got the KC-130J in 2004 and 2005, we flew it exactly like the legacy aircraft. Our first operational mission was to go to Iraq, and we did our missions in Iraq exactly as we did with the legacy aircraft. We did air refueling and moved cargo and personnel around Iraq and Kuwait. That was how we started.*

As we gained more experience in the KC-130J, we started evolving that and understanding the whole capability. We started operating at night with NVGs and we had to learn how to do that from scratch. And then from there we just incrementally kept expanding the envelope, adding missions as we gained confidence.

Until we've arrived at the point we're at today where, I'm not going to say we're anywhere near close to exploiting the full capabilities of the J, but widely different from where we were in 2004.

We are now doing long-range refueling of MV-22s over the Atlantic; we have introduced Harvest Hawk, both of which are game changers for us. With Harvest Hawk we have an ordinance delivering capability and we have a sensor.

Nobody in the C-130 community had any experience with that. But we went out to other communities asking for help, developing our expertise, and we have able to add those to our mission sets, and now we are at the point we have no external assistance helping us with the Harvest Hawk mission.

KC-130J configured as Harvest Hawk at 2nd MAW.

Question: How would you describe the impact of the KC-130J-Osprey pairing on USMC operations?

Lieutenant Colonel Koltick: *The two obvious changes are in your time and in your distance. Now we can operate much further and we can get there much faster when we pair the V-22 with the KC-130J compared to a traditional Marine force.*

The range that you get when you combine those two with the refueling and the situational awareness that the K/J has enabled and to be able to push that to the MV-22s, changes the context. It's hard to imagine now an environment on our mission where a MAGTF commander can't deploy an assault force.

For instance, with Special Purpose MAGTF, they're routinely thinking in terms of 1,000 nautical mile movements. Whereas a little over 10 years ago, when General Mattis took us into Afghanistan after 9/11, Task Force 58 moved 400 nautical miles and that was groundbreaking. That was a big deal, but with CH-53s and other helos, we had to land in Pakistan to do that insertion. Now, a MAGTF commander can easily plan to fly 1,000 nautical miles with an assault package.

Question: There are three KC-130J squadrons and with the introduction of the Osprey and the re-configuration of the assault force, can the fleet stand up to the demand? Or put another way, is demand outpacing supply?

Lieutenant Colonel Koltick: *The demand is high. This means that we do not have enough assets available for training with the other parts of the Marine Corps, ranging from low end to high-end training. We never have enough assets for force training.*

The second issue is that the current operational posture of the Marine Corps – forward deployed, disbursed, and facing uncertain challenges in the near and middle term – creates a demand signal that is outstripping supply and we are only in the process of finishing the currently planned full procurement of KC-130Js for the USMC.

Question: Put in other terms, the reinvention of operational capabilities has driven up the demand for KC-130Js or put in another way, the broadened concept of support, which the KC-130J can provide, has also driven up demand. Is that a fair way to put it?

Lieutenant Colonel Koltick: *It is.*

Major Pope: *One way to look at the change is to look at the transformation from Iraq to Afghanistan. In Iraq we were largely doing lift and tanking. By the time we were drawing down in Afghanistan, the KC-130J was a key element within the operations in a number of ways.*

Obviously, lift and tanking remain a bedrock of its role. But now the aircraft is part of the stack of air support and becoming in many ways a host mother ship, if you will. We are providing communications relays, doing battlefield illumination, putting mission systems aboard the aircraft because we hear the coms and then we had the Harvest Hawk strike capability as well.

Each flight of a multi-function KC-130J becomes almost a sortie generation rate functional equivalent. We can stay up 10 hours or more and what does that represent in terms of a Harrier sortie equivalent?

A major challenge facing the USMC as it leverages Ospreys and then F-35Bs will be supportable range. The KC-130J is clearly a key part of any concept of supportable range for the assault force. And with the evolving multi-mission capabilities of the KC-130J to play the role of a mothership or to provide ISR and overhead strike, the notion of supportable range can also be broadened as well

The KC-130J Demonstrates Flexibility: "Training, Training, Training"

July 24, 2014

In this interview, Major Mark Hamilton, the Operations Officer of the Squadron, talked about the significant demand on the crew to be able to provide for the flexibility demonstrated by the aircraft and essential to USMC expeditionary operations.

In the discussion with Major Hamilton, the significant demand which the KC-130J placed on training of the crew, understood as a continuous process of certification and recertification.

As Major Hamilton put it: *It is a highly flexible platform. But to maintain that flexibility requires constant training. You need to train for all the missions and be able to move rapidly from support and execution of battlefield illumination, long range tanking, air drops, or close air support in the case of Harvest Hawk.*

Preparing the Simulator for KC-130J Training at VMG-252.

He emphasized that it was a question of learning and re-certifying for each of these missions and having the ability to flip among them in order for the KC-130J to be used in a flexible manner to support the deployed force: *For example, with regard to flight codes you have 90 days to re-certify for air drop missions and 365 days for battlefield illumination.*

The Harvest Hawk mission introduced a new demand set for

the KC-130J crew, namely close air support. The community could draw upon the USMC core competencies for CAS but before Harvest Hawk this competency was not a central piece of the KC-130J mission set.

As Major Hamilton underscored: *We can provide loiter time plus speed that neither fast jets nor UAVs can provide. We give the ground commander another option, namely an asset faster than a UAV with more loiter time than a fast jet. This option obviously can operate only in a situation of air dominance, and be used in areas where it is appropriate. But it provides the Marines with another option for the ground commander.*

The KC-130J as an Enabler: The Perspectives of a Crew Chief and a Maintainer

August 7, 2014

In this interview, Sgt Paul Millis, a Crew Chief, and Sgt Thomas Chevalier, a power train maintainer highlight their perspectives on the KC-130J and its operations. Both have Afghanistan experience, and Millis has Pacific experience as well.

The discussion underscored two important themes with regard to the use of the KC-130J by the USMC.

- The first theme is the significant versatility of the aircraft by which it was used across a wide variety of missions in Afghanistan and is a key enabler of the SP-MAGTF.
- The second theme, which reinforced the first, was that the close integration of the crew with the maintainers was crucial to the rapid turn around characterized by Afghan operations.

Sgt Millis underscored that in Afghanistan each KC-130J was used in excess of 100 hours per month. The crews had to perform rapid turn around for missions in which the mission would change 4-5 times during the day and the plane needed to prepare for the new mission.

Rapid taskings were a norm of the Afghan operational tempo. Among these missions were air refueling, passenger and cargo lift, precision airdrops, battlefield illumination, rapid ground refueling and overhead strike and reconnaissance via the Harvest Hawk.

Sgt Millis: *We were constantly reconfiguring for missions based on the demand. And the demand for battlefield illumination at night required a rapid preparation of the aircraft for support to the Marines on the ground.*

Sgt Millis highlighted the advantages of having a loadmaster separate from a crew chief, which was a key element enabling rapid tasking requirements. This system is being replaced by a Crew Master system in which the Crew Master is asked to do both tasks.

Sgt Paul Millis, a Crew Chief, and Sgt Thomas Chevalier, a power train maintainer for the KC-130J.

The KC-130J was used extensively to provide supplies so that IED riddled roads could be avoided and this meant that lives were saved by the KC-130J mission.

Sgt Millis also highlighted the key role of precision air dropping: *We were able to drop 13-14 bundles into an area the size of a football field and we could prepare this load and deliver it within a five hour turn around period.*

Sgt Chevalier was in Afghanistan as the Harvest Hawk was introduced and this provided a challenge but because of the close integration and working relationship between maintainers and the crew the challenge was met. He noted: *Maintainers on occasion flew with crew so they could see the results of the Harvest Hawk capability. We came back knowing we saved lives and understood that our job served a higher purpose.*

Sgt Chevalier highlighted that the durability of the KC-130Js really enabled the SP-MAGTF because the Ospreys needed to be supported in order to operate with the reach and range desired: *It is a question of sustainable range and the KC-130Js provided that sustainable part.*

The interview with the two young Marines highlighted the importance of the integration of the crew with the aircraft to deliver the results needed in a difficult operational theater such as Afghanistan. And this teamwork is going forward to support whatever is next for the KC-130J in its enabling role for the MAGTF.

Operating the Harvest Hawk: Shifting the Operational Context and Next Steps

July 31, 2014

I spent a morning in the KC-130J simulator watching Marine pilots hone their skills and did so from the tanker seat in the aircraft.

After the simulation experience, I sat down with one of the pilots, Captain Michael Jordan, to talk about his experience working with the Harvest Hawk in Afghanistan. The Captain is of the new generation of USMC pilots who have flown the KC-130J from the beginning and so the Harvest Hawk experience seems a "normal" evolution and simply preparing for the next transition, whereby the "mothership" can handle data, C2 or ordinance dependent on the evolution of USMC concepts of operations.

Jordan noted: *The Harvest Hawk was first introduced into Afghanistan by 3rd MAW so the squadron went to the West Coast to learn their skills in preparation for their use of Harvest Hawk in Afghanistan in 2012. We trained as Harvest Hawk co-pilots right before we deployed for Afghanistan last year. And I would say that at least half of the hours we flew in Afghanistan starting in July 2013 through February of this year were Harvest Hawk missions. After that I went to Spain and became part of the Special Purpose MAGTF.*

He emphasized as well that it was different type of flying than on other KC-130J missions: *A lot of what you are doing is working in orbits, circling around targets, talking with people on the ground and managing the battlespace. 95% of what you are doing is flying, holding your orbit and then waiting for that 5% of the mission where you prepare to and then launch your ordinance.*

And in flying the plane the operation from the cockpit is different as well. With the Heads up Display you are flying the aircraft. Now you need to not just focus on just flying the aircraft and executing the mission but also work with the navigation radar to see where you are going and what you are doing within the battlespace.

Captain Michael Jordan, VMGR-252.

Question: What are the trade-offs necessary to operate a KC-130J as a Harvest Hawk?

Captain Jordan: *We lose the external tank on the left side as a sensor is placed on that tank. We can not fuel from the tank. This means that we reduce total fuel capacity.*

A typical J would have a max fuel capacity of 60,000 pounds dependent upon fuel conditions. With the Harvest Hawk configuration we can have a max fuel capacity of around 42,000 pounds. This reduces our time on station.

The Marine Corps is looking at an upgrade which would put the sensor on the front of the plane rather than the external tank and would recover the use of the tank.

The other aspect is that we lose the left side Aerial Refueling pod. We have four hard points on the left side for the missiles; we can do refueling but with only one hose on the right hand side. We can not have two houses serving the tanker mission.

Question: How are operations different from inside the aircraft to operate the Harvest Hawk mission?

Captain Jordan: *It is quite different. In Harvest Hawk the pilot becomes more of a battle manager, while the co-pilot flies the plane. In the back of the aircraft we have a different crew as well. We generally have two officers in the back; a Fire Controls Officer sits in the right seat and operates the sensor to provide for target acquisition. Typically, we have an assistant in the left seat to aid with the communications traffic and to assist the FCO.*

Harvest Hawk system on KC-130J at 2nd MAW.

Question: Where does Harvest Hawk go next?

Captain Jordan: *The entire Harvest Hawk experience highlights the utility of a "mother ship" in an air dominance environment. There is no reason that we cannot take data from UAVs or the F-35s or the Harrier litening pods and be able to contribute to combat management or support to the ground commanders.*

The Challenge of Sustaining the ARG-MEU in an Era of Distributed Operations

March 30, 2014

During the visit to New River on February 10, 2014, there was the opportunity to sit down with LtCol Boniface to discuss his most recent experience as the ACE Commander of the 26th MEU, which has just returned from the Middle East..

A key focus of discussions with Boniface in the past and in the current one has been with regard to shifting a helo centric ARG-MEU thinking in terms of a 200 mile box of operations to one transformed by the Osprey into operating over a 1000 miles.

This has led to the deployment of three ship formations no longer within a 200-mile area but over more than a 1000-mile area. This leads in turn to a major challenge of re-supplying the ARG-MEU. And the problem seen here is being replicated by the USMC need to pursue a distributed laydown strategy in the Pacific.

Sustainability over distance is a key challenge as the geography covered expands and the ACE assets can operate over those greater distances as well.

LtCol Boniface highlighted two key challenges.

- The first is simply the challenge to the Military Sealift Command to support a disbursed ARG-MEU.
- The second is having a responsive and effective parts availability pool to support the deployed but dispersed ARG-MEU. This is an especially important challenge for the Osprey because of relatively limited locations within which parts are available to be flown or delivered to the ARG-MEU on deployment.

Put another way, the deployment of the ARG-MEU is not constrained by Osprey operations, but the effectiveness of the logistics or sustainment operations. The carriers get supplied every week; the ARG-MEUs only every 10-14 days. This disparity no longer makes sense given the reality of ARG-MEU operations under the influence of the Osprey. In effect, there is a tactical limitation posed by sustainment which can have strategic consequences.

An additional challenge is the load being placed on the USMC Ospreys by resupply to the Navy ships.

LtCol Boniface: *About 20 percent of my flight hours on the MEU were used to help the ARG-MEU to do resupply. We need to get into a different way of thinking to release the Ospreys from this mission. We need to think through a different rhythm or approach to sustainment from the at sea replenishment system.*

Comment: You were talking about the CV versus the MEU. The MEU historically has been really defined by the box that you described. It's a helo-defined box, more or less. And, and so as the Osprey now you can operate at a much wider circumference and is much more useful to the joint force with the special purpose MAGTF people are interested from the joint force and the Navy itself is looking at Ospreys for resupplying. What that tends to mean is that your hermetically sealed Gator Navy makes no sense.

Marines and Sailors assigned to Maritime Raid Force, 26th Marine Expeditionary Unit (MEU), embark from the USS Kearsarge (LHD 3), at sea, on MV-22B Ospreys assigned to Marine Medium Tiltrotor Squadron (VMM) 266 (Reinforced), for a simulated night raid, Feb. 09, 2013. (U.S. Marine Corps photo by Cpl Kyle N. Runnels/Released).

LtCol Boniface: *That is exactly what I'm saying. And as we transform the ARG-MEU with the F-35B, the CH-53K, the new Cobra and Huey, we need to significantly rethink the logistics support structure for the deployed fleet. We have gotten out of the helo mindset with regard to operations but not with regard to sustainment approaches.*

Another problem identified by LtCol Boniface is the constraint of having very few supply locations from which Osprey parts can be generated to the deployed force.

LtCol Boniface: *We have gotten to the point where it is a very reliable aircraft, but it needs parts to fly. We are at a transition point whereby the flow of parts to the fleet needs to be significantly improved from what will need to become a global supply chain to the aircraft which is being deployed widely now in the Pacific, Europe and the Middle East, and will now be bought by several allies. This really raises the benchmark on global supply of parts to the aircraft on deployment.*

It's not about the plane anymore. It's about the entire logistics enterprise that sustains operational effectiveness

The MV-22 and the MEDEVAC Mission: Thinking Through the Capability in Afghanistan

March 26, 2014

During the visit to New River on February 10, 2014, LtCol Ennis discussed his experience with the Osprey, and preparing for operational innovations, such as CASEVAC. Also discussed was his preparation for working with the CH-53K program.

LtCol Ennis has been with the Osprey program for a considerable period of time, from 2003. He was part of VMX-22 and worked with the introduction of the Osprey into USMC operations. He is becoming the government's flight test director. He will be bringing his Osprey experience to the CH-53K test program and he noted that one area where the MV-22 is very good and the CH-53 is not is with regard to brownout situations.

The Osprey has a good capability to land in a brownout situation whereas the CH-53Es currently would rather encounter the enemy than deal with a brownout situation. They would rather land right next to a compound that has a known enemy as opposed to land in open desert.

He also described his experience of going from test pilot to operational experience in Afghanistan and its importance to the evolution of the aircraft itself. They are a lot of things the test community is working on that are not necessarily the focus of the actual fleet which is deployed. This gap needs to be closed.

During his time in Afghanistan last year, he was involved in shaping a MEDEVAC role for the Osprey. Clearly, the advantage of the Osprey over current Army rotorcraft is their ability to operate over a much broader range without the significant Forward Operating Base infrastructure required by the Army medevac approach.

LtCol Ennis During SLD Interview at New River, February 10, 2014.

He described that in a situation where there are many troops and FOBs, the current U.S. Army approach works well and operates largely in a 40 nautical mile radius. When forces are dispersed and one does not have a large support infrastructure, the situation is different and the Osprey can perform the MEDEVAC mission over a much larger area without a significant FOB infrastructure.

He underscored: *We put Ospreys in two different locations to cover an* operating area. *The two can cover the area in between if they pick up from one location and drop to the other. We went out to something like 160 nautical miles because if you can fly between point A and point B, it is basically a big ellipse which works out to a distance of 160 nautical miles.*

The Osprey can be fitted nicely with the MEDEVAC gear as well. If you put every litter in the back it can hold 12 patients, but that is not ideal. You would prefer six so that you can have the medical staff and support gear spaced out.

Normally, you would strap the gurneys to the floor, rather than use litters, because at that point you can work all around the patients. Outfitted for a MEDEVAC operation the seats fold up and the litters stack up against the walls. Notably, the USMC prepared for the possible use of the Osprey in the MEDEVAC role for the very beginning of its deployment history.

Now, the possibility should become a reality to transform the capabilities for MEDEVAC operations intra-theater.

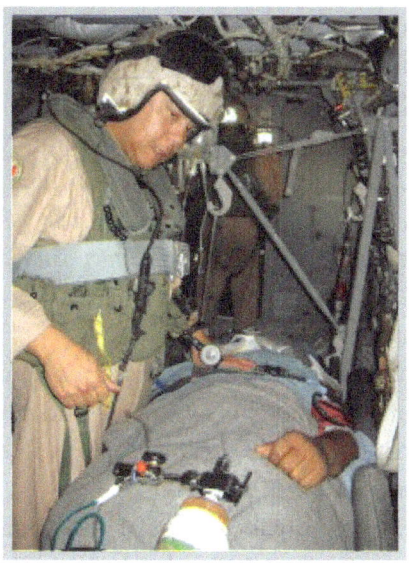

Chief Petty Officer Richard B. Guerrero, a Chief Hospital Corpsmen with Marine Medium Tiltrotor Squadron 263 (Reinforced), 22nd Marine Expeditionary Unit, cares for a patient aboard an MV-22B Osprey during an emergency medical evacuation June 25, 2009. (Official USMC photo)

Visiting a USMC Unmanned Aerial Vehicle Squadron: The VMU-2 Discusses the Future

July 15, 2014

It was a pleasure when visiting the Second Marine Air Wing in June 2014, to be able to visit with the VMU-2 squadron and to get a perspective on how UAVs are part of the future of combat, and not learn that they are the future of combat.

Clearly, the squadron is looking at the mix of airborne support elements to the MAGTF, the F-35B, the legacy fighters, Ospreys, rotorcraft, the KC-130J and UAVs, and working on how to shape the UAV contribution within an evolving context.

The CO of VMU-2 is Lieutenant Colonel Kris Faught. The CO most recently served with VMM-266 as the Air Combat Element Operations Officer for the 26th MEU. The discussion was wide-ranging and focused on the various dynamics of change affecting the evolution of UAVs within the USMC and their potential role in operational support.

The Mission Statement for the squadron highlighted its contextual support role, both now and evolving future capabilities. The Mission of VMU-2 is described as follows: *To support the MAGTF Commander by conducting electromagnetic spectrum warfare, multi-sensor imagery reconnaissance, combined arms coordination and control, and destroying targets, day or night, under all-weather conditions, during expeditionary, joint and combined operations.*[4]

The CO highlighted some key limitations facing UAVs for the USMC, and identified what he saw as a solid growth path going forward. He started with a discussion of the value of the Shadow UAV and its experience for the USMC.

Clearly, Shadow has provided important operational experience, yet the Shadow is not congruent with where the USMC is headed. As he put it: *It is clearly not expeditionary and looks like it has been designed by a tanker.*

This has meant that the USMC is looking for a more agile UAV with a significantly reduced footprint to support its operations.

Currently, the squadron is working with the RQ-21A Blackjack to work through ways to evolve an expeditionary support capacity from its UAVs.

Notably, 2nd MAW Forward is using working with an Early Operational Capability (EOC) run up to the RQ-21A in Afghanistan to gain operational experience in order to help shape the way ahead for the UAV role within the USMC.

The principle differences are software and some ship compatibility issues with launcher and retriever between the EOC system and the RQ-21A system.

In the slides provided in the briefing by Lt. Col. Faught, a number of key characteristics of the RQ-21A were highlighted:

- Early Operational Capability.
- Max Control Range of 93 kms or 50 nms.
- Service Ceiling of 15,000 ft MSL
- Airspeed of between 55 and 110 knots.
- An endurance of up to 15 hours.
- Operates with a low noise signature.

The RQ-21 Blackjack as seen at USMC Air Station, Cherry Point, North Carolina.

A key quality of the Blackjack is its non-proprietary payload system. He underscored: *The payload bay is not patent-protected. This means that L-3 is building payloads. Lockheed Martin is building payloads. Little one off shops in San Diego are building payloads.*

And clearly the trend line, which the Marines would like to see, is an ability to shape modular payloads to provide for the support missions envisaged for UAVs.

Currently, the Blackjack carries the following types of payloads:

- Electric-Optical.
- Mid-Wave Infrared.
- IR Marker.
- And the Secondary Payload Bay Supports: CRP and EW.

The approach to support the RQ-21A is considerably less than the Shadow and clearly allows the Marines to work on their expeditionary support approach. With regard to the RQ-21A, the squadron was working with industry to shape ways to enhance capability.

The RQ-21 Blackjack as seen at USMC Air Station, Cherry Point, North Carolina:

LtCol Faught highlighted the following: *We are looking at size, weight, power tradeoffs to enhance overall platform capability. Currently, we are at 135 pounds with the platform and we could go as high as 165 which would give us more payload to carry onboard.*

LtCol Faught emphasized throughout the discussion the need to evolve the payloads along with other key aviation capabilities being shaped for the MAGTF. He especially felt that EW payloads will be increasingly of interest going forward. And he felt as the F-35B joins the force along with the Ospreys, the opportunity to rework evolving UAVs to operate with these more expensive combat systems would be significant.

An ability to command UAVs from Ospreys is significant and the UAV can be launched in advance of Ospreys entering an objective area and providing a wingman or escort function for the Ospreys. The same could prove true for the F-35Bs dependent upon the mission set.

The core point is that the squadron is not thinking as an island. But there is a core challenge facing the squadron in terms of shaping an expeditionary approach. The support equipment is provided by different parts of the USMC, whether they be trucks, energy support or C2 and these different parts are not bought with the UAV in mind. Such an approach clearly will not maximize the expeditionary footprint for the UAV.

LtCol Faught also focused on shipboard support by UAVs. Clearly, one way to think about this is flying the son-of-Scan Eagle, which is the RQ-21 off of ships, for it has been designed to be recovered shipboard. But he felt that support to an amphibious fleet could come from large land-based UAVs as well.

We could launch an UAV from North Carolina or Europe and it could cover our area of interest in Africa, for example. It would be providing shipboard support but does not need to be launched from the ship. We need to think through operational concepts to come up with the most effective and sensible approach.

He emphasized as well some of the misconceptions about UAVs.

One of these is cost. *Perhaps the per unit cost looks good but if you lose a lot of them, the cumulative cost does not look so good.*

And he argued that the UAVs are really not good airplanes: *If you asked a child what that thing on the tarmac is and he would say it is an airplane. It is an airplane. But it is not built like an airplane. Indeed, they are built with unique and often not so reliable parts. It would be good in the future if* we could tap into a broader aviation construction industry to get parts for future UASs or UAVs.

There are other limitations as well affecting training and operations: *Amazingly waterproofing is a problem with the RQ-21. And climb rates are difficult which means that you want to operate in relatively flat terrain*

A particular vexing problem facing UAV operators in the United States is to be able to operate UAVs in civil airspace. The FAA has been mandated by Congress to be able to do so by 2015.

The squadron has built a capability to do just that and recently a NASA team came to Cherry Point to see how the Marines carried this off. I was able to visit the control room where the Air Traffic Controllers managing the flight of USMC UAVs into the civil airspace in their region.

Sgt Lopez and LtCol Faught standing in front of the RQ-21A.

They told me: *We have a ground-based sense and avoid system for our UAVs here at Cherry Point. The only other place in the U.S. where you can find such a system is at Cannon Air Force Base in New Mexico.*

I then visited the hanger and the RQ-21A on the tarmac. I had a briefing from the pilot, Sergeant Lopez who explained the challenge of flying an UAV. It was clearly a very cool digital experience, but guiding an out of sight airplane has its challenges.

When asked what was the next challenge he was looking

forward to with the program, the answer came quickly: *I want to be the first at-sea pilot to operate the aircraft!*

A final issue, which we discussed with a Prowler pilot in the room, was the whole challenge of transitioning the Prowler experience into the UAV squadron and the F-35B squadron.

As Lieutenant Colonel Faught put it: *We need to find ways to exploit the analytical infrastructure which has supported Prowler and take that forward into the 21st century approaches we are now shaping.*

Ospreys and Landing Zone Flexibilities

February 11, 2014

The operational flexibility of a tiltrotor is clearly seen when the Marines approach landing sites. The ability to maneuver through the battlespace to shape alternative pathways to an LZ is important. The ability to get away rapidly in transition from rotor to plane mode is also a key capability.

The flexibility generated by the nacelles is a key facilitator of the Osprey's capability.

The process of rotating the nacelles between helicopter and airplane mode is called "transition", and the reverse from airplane mode to helicopter mode, "conversion". Transition and conversion procedures are simple, straightforward, and easy to accomplish. The amount and rate of nacelle tilt can be manually controlled by the pilot or can be performed automatically by the flight control system.

The V-22 can perform a complete transition from helicopter mode to airplane mode in as little as 16 seconds. Conversions and transitions can be continuous, stopped partway through, or reversed as desired. A tiltrotor can fly at any degree of nacelle tilt within the authorized conversion corridor envelope.

During vertical takeoff, the conventional helicopter controls are utilized. As the tiltrotor gains forward speed, the wing begins to produce lift and the ailerons, elevators, and rudders become more effective.

Between 40 and 80 knots, the rotary-wing controls begin to be phased out by the flight control system. Once in airplane mode, the wing is fully-effective and pilot control of cyclic pitch of the proprotors is locked out. Because the nacelle angle can be commanded separately from the primary pitch controls of rotor cyclic and tail elevator, the conversion corridor (the range of permissible airspeeds for each angle of nacelle tilt) is very wide (about 100 knots).

In both accelerating and decelerating flight, this wide corridor means that a tiltrotor can have a safe and comfortable transition or conversion, offering the combined advantages of speed and maneuverability for low level flight.[5]

During a visit to 2nd Marine Air Wing in the winter of 2014, Murielle Delaporte and I visited VMX-1 and met with Col Orr. During the visit, Marines were training their variant ingress and egress tactics with the Osprey and we rode along. Murielle Delaporte provided a photo after disembarking, and I was riding on the Osprey to the right commanded by Col Orr. Credit Photo: Murielle Delaporte.

During a visit to New River Air Station in North Carolina on February 10, 2014, we were (Murielle Delaporte and myself) able during a USMC training session to experience the ability of the Osprey to land and depart LZs rapidly and the transition and get away speed of the airplane mode.

This flexibility is a core combat capability provided to enable the Marines getting off and getting back onto the plane enhanced security and effectiveness.

Not always easy on the stomach, and it would be better to be in the front of the aircraft, when such flexibility is demonstrated, but the Osprey is clearly not a helicopter when it comes to the LZ.

VMX-22 Aboard USS AMERICA: An Interview with the CO of VMX-22

November 8, 2014

VMX-22 flew to USS AMERICA and worked with the crew in shaping an initial maintenance regime aboard the ship, providing crucial insights into how the new amphibious assault ship enhances Marine Corps aviation capabilities.

The ship features a hangar bay below the flight deck where maintenance is performed, unlike legacy large deck amphibs. The USS AMERICA class ship has a much larger hangar bay conducive to conducting maintenance that otherwise would have to be conducted on the flight deck..

This means when weather gets dicey, aircraft can be put below deck to protect them from natural elements, avoiding the risk of weather damage compared to remaining on the flight deck on legacy amphibs. The ship has three synergistic decks which work together to support flight deck operations, with maintenance done in a hangar bay below the flight deck rather than topside like traditional amphibious ships.

In my interview with the ship's Captain Robert Hall highlighted key capabilities: *The ship has several capabilities which allow us to stay on station longer than a traditional LHA and to much better support the Ospreys and the F-35Bs. These aircraft are larger than their predecessors and need more space for maintenance.*[6]

Captain Hall noted the ship has two high-hat areas to support maintenance, significantly greater capacity to store spare parts, ordnance and fuel, and can carry more than twice as much JP-5 than a traditional LHA. *By removing the well deck, we have a hangar deck with significant capacity to both repair aircraft and to move them to the flight deck to enhance ops tempo*, Hall explained.

In an interview with VMX-22 CO Col Robert L. "Horse" Rauenhorst, the commander provided insights on the USS AMERICA experience.

Maintenance wise, the hangar bay provided two dedicated spots to perform maintenance with the wings spread, as compared to legacy amphib ships with only one dedicated spot, Col Rauenhorst said. *When we encountered heavy sea states, freezing rain and snow around the Straits of Magellan, we were able to hangar all 11 aircraft, including four CH-46s from HMM-364 and three SH-60s from HSC-21, to protect them from the elements.*

VMX-22 operated aboard the USS America working on ship integration with the Osprey.
Credit: USS America

Once clear of weather, the Air Boss could quickly return aircraft to the flight deck and resume normal operations that otherwise would have been delayed in de-icing. *The Marines onboard were really impressed with the size of the hangar bay and being able to do maintenance in protected spaces, rather than on the flight deck as before.*

The fuel capacity onboard far exceeds legacy amphib ships. *USS AMERICA has the capacity to carry a lot more JP fuel compared to legacy amphibs with a well deck. With the efficiencies of the USS AMERICA's generators and not operating off turbines, the ship operated at 12-13 knots while conducting flight operations, ultimately saving money in operational efficiency terms.*

Working with Captain Hall proved highly successful. *Bob was really good to work with. He was really excited to have Marines come onboard because he had never deployed with Marines before. It was a very good working relationship.*

The crew, including veterans and newcomers, showed exceptional motivation. *The crew was super motivated to get the aircraft onboard and operate it. We did crawl, walk and run phases settling into flight operations, starting with day and night Carrier Qualifications off Pensacola. By deployment's end, Captain Hall had a seasoned crew doing both tilt rotor and rotary-wing day and night operations without flaw.*

The deployment marked the first time MV-22s operated in most visited countries including Columbia, Trinidad and Tobago, Brazil, Uruguay, Chile, Peru, and El Salvador. *The reaction was of real interest, and as one person put it, 'it is like an airplane out of the movie Transformers!' referring to the MV-22 transitions from helicopter to airplane mode.*

Senior regional leaders showed particular interest in the time/distance equation and the aircraft's ability to operate in difficult terrain. *The Brazilians were very interested because of their country's scope and the need to operate into the Amazon region. It takes them a long time to get logistics resupply and fuel out there using helicopters.*

Operationally, the Ospreys could fly Theater Security Cooperation teams 350 miles prior to the ship's arrival, giving Marines from Special Purpose Marine Ground Task Force-South an additional day on both sides of USS AMERICA's visits for building partner capacity and conducting subject matter exchanges.

When asked about the tiltrotor-enabled USS AMERICA's advantages as an assault ship, Col Rauenhorst emphasized flexibility: *When you get into that 1200-mile range or less, and combine with the organic capability of KC-130s, you have a powerful moving forward-operating base that can extend the operational reach for the Marine Corps.*

VMX-22 was scheduled to receive four F-35Bs by summer 2014 at Edwards AFB, eventually expanding to six when the detachment moved to Yuma. The unit was supporting digital interoperability across all aircraft types including Unmanned Aerial Systems and Marine Air Command and Control Systems.

The priorities of Lieutenant General Davis, the DCA are current readiness, future readiness, and digital interoperability, Col Rauenhorst explained. *We're working on how to integrate and share information with F-35, MV-22, CH-53, UH-1, AH-1, MQ-21, and A/NTPS-80 through a common opera-*

tional picture while supporting Marines on the ground with real-time information.

The focus includes experimentation, tactical demonstrations, and operational testing of digital interoperability systems. *We're basically refining how we share and move information not only amongst aviation platforms, but determining how to get that information to Marines on their tablets, whether in the back of an assault support aircraft or in the objective area, so everyone works from a common operational picture with increased situational awareness.*

TEN

2015 Visit: Strands of Transition

My 2015 visit provided a snapshot of the Marine Corps during a significant transitional period as they were adapting to new technologies, operational concepts, and strategic priorities following the engagements in Afghanistan and Iraq wars.

The articles in this chapter capture a Marine Corps in transition, with several major themes emerging:

- Transition from COIN to Hybrid Warfare.
- Naval Integration and "Return to the Sea".
- New Aviation Platforms Transforming Operations.
- Electronic Warfare Integration.
- Command and Control Evolution.
- Unmanned Systems Integration.
- International Cooperation.

July 10, 2015
By Robbin Laird and Murielle Delaporte
During our visit to 2nd MAW on May 18, 2015, we had a

chance to talk with two members of VMAQ-3, an electronic warfare squadron flying EA-6B Prowler jets. It is part of Marine Aircraft Group 14 under the 2nd MAW.

With the first Prowler pilots being trained this summer at Beaufort for the F-35B, a transition is already underway. The Prowler is due to be retired in 2019 and will be replaced by a wide-ranging focus on EW throughout the MAGTF.

Clearly, unmanned aircraft or remotely piloted vehicles will be added to the mix as well. Currently, the USMC Blackjacks carry payloads to contribute to EW and over time the UAV element will enhance its role as well.

In an article by Joshua Stewart in the *Marine Corps Times* published on January 29, 2015, the shift in roles of Marines in supporting EW was discussed.

As the EA-6B Prowler flies into retirement and the Corps takes a new approach to electronic warfare, some Marines who spent their careers in the radar-jamming aircraft will be transferred to other military occupation specialties, and many will work with unmanned aircraft.

The RQ-21 Blackjack as seen at USMC Air Station, Cherry Point, North Carolina:

About 10 percent of Marines in the 7588 electronic warfare officer MOS will become 7315 unmanned aircraft systems officers. Concurrently, the Corps is changing the duties of the 7315 MOS.

"The new 7315 MOS will provide us with a cadre of better-trained and more versatile [unmanned aircraft systems] officers capable of serving in a variety

of operational roles in support of Marine Corps doctrine," said Maj. Paul Greenberg, a Marine spokesman at the Pentagon.

Most of the transitioning electronic warfare Marines will be company-grade officers with a primary 7588 MOS, Greenberg said. Other electronic warfare officers will stay in that MOS until they leave the service, and will serve in B-billets, he said.

The electronic warfare community's transition is one facet of the Corps' new approach to controlling the electromagnetic spectrum in battle, and the service's new philosophy is more complex than merely rolling dozens of Marines from a niche community into a new MOS.

Currently, Prowlers are the service's electronic warfare workhorse, but in the future, a variety of platforms — including unmanned systems, rotary aircraft and ground vehicles — will also be a part of this warfare domain. It amounts to a Corps-wide makeover of electronic warfare in which manipulating and monitoring the electromagnetic spectrum is a more integral part of every aspect of combat.[1]

The mission of the squadron VMAQ-3 is identified as follows:

Support the Marine Air-Ground Task Force (MAGTF) commander by conducting airborne electronic warfare, day or night, under all weather conditions during Expeditionary, Joint, or Combined operations.

We interviewed two members of the squadron who recently joined the EW world. Captain Casey Jacobs, Prowler pilot, has been doing this job for over three years. This is his first fleet squadron. Captain Colton Browser is an electronic weapons officer and he been in the squadron for just under three years as this is his first tour as well.

According to Captain Jacobs: *One of the challenges which the Prowler community has is that it has more in common with other EW communities in the Air Force and Navy than it does with many Marines. This will change as EW becomes a ubiquitous part of USMC operations. The feedback we get from Marines is that the plane works in supporting them. But there is a lack of general understanding of what we do, and the problem lies in that there isn't a wide understanding of what we do and how we contribute. We work a lot with the USAF as well. And as such we are part of general support for the leadership with the Combined Air Operations Center during operations.*

Captain Browser added that the Prowler is understood more by myth than reality: *We go to airshows and people have understanding based on myths more than realities and, for example, assume that our fuel probe is a laser beam and that sort of thing.*

It is quite different when one goes to the Weapons Tactics and Training Courses at MAWTS where integration of EW within the combat capability of the USMC is a key focus.

As Captain Browser put it: *At WTI there is a clear focus on how to integrate EW into combat operations. At our most recent WTI, we participated in all of the various air support tasks for the MAGTF, including air to air and close air support. This is in contrast to our Red Flag experience where we are treated as a specialized asset in the battlespace.*

And with the coming of the F-35, MAWTS is working on how to integrate the plane with its various capabilities into the force, and that includes EW. Some Prowler pilots will transition to the F-35 community as well.

The members of the squadron noted that there are four Prowler squadrons overall with three of them at Cherry Point with 18 planes per squadron. The Prowlers are a high demand asset and in constant rotation to combat theaters where they are attached to different commands for whatever they are needed for.

As Captain Browser put it: *We are focused on the entire electromagnetic spectrum, so anything from communication, to RF energy produced by radars etc. It is a very large spectrum that we need to contain to have combat superiority. The Prowler has limitations on what we can control within that spectrum, but going forward the USMC approach will be that not just one platform can do everything. We are going to shape a collaborative approach to achieve spectrum dominance.*

The Marines are blending the various MOSs or Military Operational Specialists in shaping a forward leaning electronic warfare approach.

According to the article by Stewart quoted earlier: *The MOS transition process will start as early as October as Prowlers begin to be phased out, Greenberg said. The transition will continue through fiscal 2019 when the aircraft leaves the fleet.*

About 80 percent of electronic warfare officer billets are expected to become unmanned aircraft systems officer billets, he said. Other electronic warfare officers are expected to transition to other pilot or naval flight officer MOSs, or switch into intelligence or communication communities.

Greenberg said others will join the Marine Air Ground Task Force Cyberspace and Electronic Warfare Coordination Cell, a unit that will help the Corps develop its cyber and electronic warfare capabilities.[2]

For Captain Jacobs: *The Navy has tended to be more stove-piped than the USMC with regard to EW. What we have had to do to survive and thrive is to work with the infantry to provide EW directly to them. They may not understand completely the process but they like the product.*

This means that a broader range of MOSs than simply EW ones are looking at cyber and EW and what it contributes to the success of their own combat area of specialized interest. This is creating a more integrated approach.

In other words, the mental furniture of various combat specialties is being recast to open the aperture to understand that electronic warfare is not a specialty but a constant of combat success.

This has been a consistent theme within Marine aviation since the decision was made to sundown the Prowlers in 2009. At the time, LtGen Trautman, then Deputy Commandant for Aviation, said:

Employing a manned low density, high demand electronic warfare platform makes absolutely no sense in a world that will be dominated by F-35, unmanned aerial systems, and full digital interoperability among every aircraft and sensor operating within the future battle space.

Electronic warfare aircraft like the Prowler and Growler are, by their nature, self-limiting technologies that are an anachronism of the last century.

We can and must do better.

Preparing to Operate Off of the HMS Queen Elizabeth: Working with the Marines at VFMAT-501

June 10, 2015

By Robbin Laird and Murielle Delaporte

During a visit to the 31st Marine Air Group, we visited VFMAT-501, the Warlords and met with the CO of the Squadron "OD" Bachman, Major Brian Bann and Squadron Leader Hugh Nichols from the Royal Air Force.

The visit to the training squadron as well as to the USS WASP the following week drove home a core point – the Brits and Marines are working closely together to stand up their separate but coordinated capabilities associated with an F-35 enabled 21st century combat force.

The F-35 global enterprise is a key enabler of the use of collaborative resources. The Brits are training at Beaufort on F-35 equipment at the base, including the simulators, as their own facilities are stood up in the UK and the squadron grows before returning to the UK to get ready to work with the HMS Queen Elizabeth.

The Brits are integrated members of the squadron and the Marine Corps and British maintainers are learning together how to adapt their specific maintenance protocols, which are different, to a common airplane.

Obviously, this will play real dividends down the road in terms of being able to cross deploy at sea. And the Brits recognized that a software upgradeable airplane requires continuous upgrade in order to stay at the leading edge. They are keeping a permanent detachment at Edwards AFB to remain engaged in the lifetime modernization envisaged for the F-35 global fleet.

Question: What is your function here at the squadron?

Sqn Ldr Hugh Nichols: *I have two roles. I am an instructor pilot within the Warlords and in that role, I am an integrated member of the team. My other role is as the Senior National Representative for the UK on the base here.*

Question: At Luke the Aussies and USAF pilots are flying each other's planes. Is that happening here?

Sqn Ldr Hugh Nichols: *It is. In effect, we have a pooling agreement here. Our aircraft are pooled with those of the Marines, and we fly aircraft in the pool not just the UK jet.*

Question: When you return to the UK with the planes, obviously a wider F-35 community is being established with which you will operate. How do you see that?

Sqn Ldr Hugh Nichols: *The majority of the operating areas big enough to fully utilize this aircraft will be out over the North Sea, so I can see us using this to our advantage by operating with our Northern European allies. I would anticipate that there will be a lot of cooperation with Norwegians,*

Danes or the Dutch as we bring this exciting aircraft into service on European soil.

Sqn Ldr Hugh Nichols during 2015 visit.

Question: And because the B and the A have common combat systems, your collaboration will not depend on which airframe you fly?

Sqn Ldr Hugh Nichols: *That is correct. At the end of the day, it doesn't matter if you in an A, a B or C, once airborne, the mission systems are the same.*

Question: What is the advantage of being here working with the Marines?

Sqn Ldr Hugh Nichols: *There are many, but let us start with their sense of urgency in getting the aircraft to Initial Operating Capability. The Marines have done a fantastic job working through previous program difficulties and have blazed a trail towards bringing this next generation capability into service.. They are Marines, and if anything gets in the way, they deal with it. Working with them will clearly ensure that we are ready for the Queen Elizabeth.*

And the pooling agreement is important in terms of cross learning. Our young maintainers are working with Marine Corps maintainers and they are learning to work through different procedures and protocols to learn how to maintain a common airplane.

Question: Obviously, this will yield operational advan-

tages later as Marines fly onto your ships and vice versa. How do you see this?

Sqn Ldr Hugh Nichols: *Obviously, deciding to do that is above my pay grade, but clearly you are right, we have cross-decked in the past and shaping commonalities from the outset will help us to so in the future. The Marines could fly jets off of the Queen Elizabeth and we off the Wasp or other ships the USMC enable for F-35B use in the future.*

Question: The RAF is in the throes of a modernization effort and necking down to a smaller type model series of aircraft across the board. How are you working the Typhoon-F-35 integration?

Sqn Ldr Hugh Nichols: *We have already started Typhoon-F-35 integration at Edwards, with the Test and Evaluation Sqns, and it shouldn't' be too long before we are involved in training exercises on the East Coast.*

Question: Secretary Wynne made the point that modernization of legacy aircraft should be taken going forward from the perspective of working with the F-35. How do you view that approach?

Sqn Ldr Hugh Nichols: *It makes sense. Each aircraft brings different strengths to the fight and we will fly them both, with the tactics will evolving over time. Software modifications will undoubtedly be required in order to get the most out of each aircraft and ensure full interoperability; take Link 16 for example, where the F-35 could put out a huge amount of information. We need to ensure that Typhoon is able to receive and display the information without overloading the pilot.*

Question: Typhoons have flown for some time with F-22s and now with F-35s. What is the impact on the Typhoon?

Sqn Ldr Hugh Nichols: *It makes the Typhoon more lethal and survivable. Today, every legacy aircraft that can fly with a Raptor clearly wishes to do so.*

But there is going to come a point where they will prefer to fly with the F-35 due to the data linking capability of the F-35 and how that capability enhances the situational awareness of all aircraft in that fight. For example, we can push information out to the legacy fleet so they know where the threats from integrated air defense platforms are and therefore they have a better understanding of where they are safe from those systems.

Question: What is the way ahead for the British presence at Beaufort?

Sqn. Ldr Hugh Nichols: *By 2018, we will have around 250 people here. Then in 2018 we will move the squadron to the UK. 617 Sqn will fly home in mid 2018. 17(R) Squadron will remain at Edwards. It is tasked to be involved in the ongoing operational tests as new software and new capabilities come online for the F-35 throughout its service life.*

MAG-31 in Transition: The Warlords at Beaufort

June 15, 2015

By Robbin Laird and Murielle Delaporte

Marine Air Group 31 is located at MCAS Beaufort, South Carolina, home to six F-18 squadrons and one F-35 squadron. The F-35 Squadron — VMFAT-501 — trains coalition and USMC F-35B pilots. When we last saw the Warlords, they were at Eglin AFB before moving to Beaufort last year.

LtCol Bachmann now commands the squadron, taking over from LtCol Berke. When Bachmann flew the 200th F-35B sortie in 2012, the aircraft had about 800 flight hours; by June 2015, it had accumulated more than 12,300 mishap-free flight hours.

The visit highlighted that the Marines are already working integration with the legacy fleet; the Beaufort squadron flies regularly with the F-18 pilots and with the USAF, notably the Georgia Air National Guard in the Savannah Sentry exercise.

Col William Lieblein, who assumed command of MAG-31 on May 20, 2013, is an F-18 pilot who learned to fly the F-35 earlier this year. *My most recent training was to learn to fly the F-35. That was exciting and really allowed me to understand how complex the aircraft is and the capabilities within the MAGTF*, he said.

With six F-18 squadrons at Beaufort, the Marines plan to begin transitioning around 2018, but will continue flying F-18s through 2030. Col William Lieblein highlighted: *I see one of our primary tasks here at MAG-*

31 because we have the F-35 squadron here with F-18 squadrons is to develop the integration between the fourth and fifth gen, Lieblein emphasized. *We send out F-18 squadrons to fly with our F-35s. They are training with us on a daily basis.*

"OD" Bachman after his 200th sortie flight at Elgin AFB.

This integration work extends to exercises with the South Carolina and Georgia National Air Guards, as well as operations with F-22s and UK Tornados from Shaw Air Force Base.

Major Brian Bann, with the F-35 Squadron for almost three years and the eighty-first pilot to fly the F-35, described the scope of integration: *We flew our F-35s with F-22s, F-15Cs, F-15Es, F-16CJs, T-38s, F-5s, and F-18s during the Savannah Sentry exercise.*

Col Lieblein stressed the importance of developing new tactics: *When you go from the F-18 to the F-35 it is a very different experience. The problem is that if you bring your F-18 tactics with you will miss the point. That will not optimize the platform.*

For the Marines, fighter integration ultimately supports the MAGTF and Marines on the ground. *It is not just fighter to fighter but support to the Marine on the ground and integration with the MAGTF. We are working on close air support and figuring how best to do that with the F-35, and we are working with MARSOC as well,* Lieblein explained.

Comparing the F-35 to the A-10, Lieblein noted: *The A-10 was great for yesterday's war and some of today's situations, but the F-35 is good for yesterday's, today's, and tomorrow's war. It doesn't matter where that ground guy is and what type of threat he's facing, the F-35 will support him.*

Major Bann highlighted the F-35's information-gathering capabilities: *The airplane's absolutely an information gatherer. It sees everything on the battlefield—IADS, lay-downs, air pictures. It's like an AWACS/river joint almost combined. The great thing is it transmits all over Link 16, pushing information to U.S. folks, coalition partners, and ground partners.*

The cockpit fusion provides exceptional situational awareness. *You don't have to do a lot of thinking about what's going on. It describes it for you. There's complete sharing of information between F-35s as we fly together, but we're working on how to optimize that with other aircraft for the most effective operations.*

When asked about the transformation since joining the Warlords, Major Bann emphasized significant progress: *It's been a great transformation. We started with 1B airplanes with very limited software capabilities, but every month we get enhanced capabilities whether radar, EW, flight characteristics. We're doing solo flights all the time now.*

The progression from basic training to complex operations has been remarkable: *We've gone from flying basically solo at Eglin three years ago, doing basic task sets, to now flying large force exercises with F-22s, F-16s, F-18s, and full integrations with our F-35s.*

The aircraft's ease of operation compared to the Harrier is notable: *This airplane is incredibly easy to fly compared to the Harrier. We won't have massive training time to fly aboard LHAs and LHDs, and won't have to focus on takeoff and landing challenges but on the mission as our primary task.*

Regarding mission flexibility, Col Lieblein explained: *The aircraft is so much more complex and capable, we have to be prepared to do anything within its capability. Because of sensor fusion, it's going to be easier to switch from mission to mission. It will take more work up front to plan, but you'll be able to switch between missions more easily because of the fusion of sensors and weapons systems.*

LtCol Bachmann, the Warlords' CO, described the challenging transition from Eglin to Beaufort. Unlike typical squadron moves, operations continued at Eglin while simultaneously ramping up at Beaufort. Bachmann flew the first F-35B to Beaufort on July 18, 2014, with the first Marine-maintained flight on September 4.

Coming to Beaufort has been crucial to moving toward IOC. We've operated the aircraft with Marine Corps maintainers and integrated the plane into our maintenance and operations approach. The readiness of our airplanes increased by almost 30% when we fully arrived at Beaufort. Being on an all-Marine base has increased our readiness.

The pilot training evolution has been substantial: *We have produced twice as capable an F-35 pilot as we used to because the airplane's capabilities have increased. Before, we taught simple takeoff and landing, navigation, formation flight, maybe small tactical pieces. Now we're teaching day and night flying, instrument conditions, and tactical missions including close-air support, armed reconnaissance, tactical intercepts with multiple airplanes, and multi-aircraft data link operations.*

Looking ahead, the squadron faces the challenge of training pilots from various legacy backgrounds while preparing for newcomers. *We will have our first Prowler pilots training here in August, and next year we will teach pilots coming straight from T-45s to the F-35 so this will be their first fighter ever.*

The transformation at MAG-31 represents a critical phase in F-35B development, demonstrating successful integration of fifth-generation capabilities with existing Marine aviation assets while building toward full operational capability.

Visiting 2nd Marine Air Wing: Shaping a 21s Century Role for the Low Altitude Air Defense Command

June 17, 2015

By Robbin Laird and Murielle Delaporte

The Marines are in the throes of transformation. The Osprey has been a major factor, as well as introducing new C2 approaches and technologies.

A core competence of the Marines is to put infantry into an objective area and to provide force protection and support for an expeditionary force. A key contributor to this role is the Low Altitude Air Defense battalion in supporting the infantry.

As threats evolve, and Marine Corps Aviation introduces new technologies, the LAAD Marines are in the throes of change as well.

The role of LAAD and the dynamics of change were discussed with the CO of 2nd LAAD at Cherry Point during a visit on May 18, 2015. We met with Lieutenant Colonel Raymond J. Placiente to discuss the command and the way ahead.

LtCol Placiente put it succinctly: *We need to shape a 21st century employment approach but we are fielding in the case of air defense a 20th century capability. The stinger missile was designed to deal with the Mig-21 and Mig-29 and it does that well. The threat is now less that kind of aircraft and now it is cruise missiles or swarming UAVs. Even with upgrade Stinger missiles, it will be challenging to deal with the new threats.*

The Marines are in the throes of shaping a 21st century system to build out its defensive capabilities. And at the heart of this effort is the Ground-Based Air Defense and Ground/Air Task Oriented Rader or G/ATOR. The G/ATOR is an expeditionary multisession radar and is intended to replace five older radar systems.

As LtCol Placiente put it: *The Marines are looking at much richer sensor-enabled defensive capability to provide the infrastructure for force protection. We need to marry the sensor rich G/ATOR radar with effective 21st century weapons.*

Clearly, these weapons can be air, sea or ground delivered as means to attack the adversary's offensive threats to the insertion force.

In this regard, directed energy weapons are a key element of the coming capabilities for the USMC, and the ability of the power system to power both G/ATOR and directed energy weapons would significantly enhance the capabilities of the expeditionary force.

LtCol Placiente underscored: *We are looking at directed energy and high energy lasers as a key component of ground-based air defense. It is just part of the future of ground-based air defense because the Marine Corps vision is a mix of both kinetic and non-kinetic means.*

It should be noted that ground based lasers have considerable potential against UAVs, a fact already demonstrated by the laser operating aboard the USS Ponce today.

An important aspect of change will be to shape the right approach to training the Marines operating the defensive systems as well. According to LtCol Placiente: *We operate two-man LAAD teams today and the team is lead by a Sergeant or more often than not, a lance corporal. They receive a digital air picture through Link 16 to their laptop and are able to act on that information.*

For us, digital interoperability is that young corporal with his two-man team being able to receive a data link and to us that data to be able to engage aircraft with the stinger missile. At the short ranges that we're engaging aircraft we can't afford to wait for engagement command. We have to engage.

LAAD Doctrine is to engage based upon air defense warning conditions, weapons control status, ROE, and positive ID. LAAD Gunners do not have to wait for an engagement command from higher provided the target falls within the engagement criteria.

Future capabilities that improve combat ID and recognition will only improve reaction time. G/ATOR will provide a powerful sensor set for the 360-degree Marine Corps Combat force, the three dimensional warrior, but sorting out the shooter side of the equation will be crucial as well. And this is a work in progress.

LtCol Placiente highlighted the role of directed energy weapons as part of the solution set. *As directed energy technology matures and the ranges of the laser get better that coupled with the situational awareness and the cueing you can get from the G/ATOR radar is going to be the most effective weapon system we have against cruise missiles, UAVs and related threats. But that depends on the amount of power you can put out from the directed energy system.*

And as these changes occur, the configuration of the LAAD battalion will shift as well. He added: *We are going to see a significant paradigm shift in how we do business. The LAAD organization and structure of today may have to be adjusted to better exploit new capabilities and address new threats.*

LtCol Placiente underscored that the nature of the approach changes as sensors become enhanced, dispersed and command systems are not linked directly to a particular sensor system.

We could move to a system whereby the warfighter is building the air situation based on multiple sensors that are all integrated and correlated. And I don't necessarily have one guy who is sitting there and directing the fire, but maybe I have one guy who is managing the fires of multiple different strike assets.

By making use of the G/ATOR sensors and a better command and control system, we need to have the person with their finger on the trigger to be certain they are engaging a hostile and not a friendly target. When you start about swarming threats, you are going to want to push down the engagement authority to the most relevant shooter.

As we came to the end of the interview, the LtCol sounded quite a bit like the Army General who was in charge of Army Missile Defense in the Pacific when he was interviewed last year – it is not just about defense, it is about effective offense.[3]

LtCol Placiente highlighted the approach: *The best air defense is to destroy the adversary on the ground before he can attack you. We want to shape the battlefield where we can destroy the runways or their control stations or the equipment via kinetic or non-kinetic means.*

And the evolution of EW means is crucial as well.

LtCol Placiente noted: *We could well build an anti-RF bubble around the MAGTF which prevent any UAVs from being effective. For this to work, we would need to mitigate the effects on ourselves.*

The Marine CO added: *The MAGTF commander will prioritize what he wants defended and that is where we will put the majority of our effort. And that is how I am going to orient and employ my capability.*

Shaping the Way Ahead for Unmanned Aerial Systems for the USMC: A Return Visit to VMU-2

June 27, 2015

During May 2015, in a return to VMU-2, the importance of shaping a more flexible UAS for USMC operations was highlighted,

The way ahead was discussed with Captain Guy Nelson, the Operations Officer, Captain Johnathan Putney, the Assistant Operations Officer, and Captain Justin Pvlischek, the Intel Officer.

The members of the squadron provided a very clear perspective on the challenges and potential ways ahead. The use of UASs in

Afghanistan was an important phase in the evolution of UAS use within the U.S. forces.

But this really is a phase and one which needs to be put into its historical place with a clear need to move on. UASs were used in a land operation with many years of infrastructure put into place, and this infrastructure which was wide ranging, expensive and significant is hardly going to be waiting for an expeditionary insertion force.

And the con-ops learned in Afghanistan clearly are a problem as well. As one squadron member put it: *The UAS controllers were more part of the intelligence system in Afghanistan than of the Marine Corps. They were an asset which plugged into the intelligence gathering system, and did not operate as we do more generally with air assets in the USMC. Normally, the airborne assets work with the ground element and share the intelligence picture in an operational context. This was the norm in Afghanistan: an external asset managed by the intelligence system rather than organic integration with the MAGTF.*

As the operations officer put it bluntly: *We are trying to burn down the whole UAV structure which the Marine Corps created in Afghanistan and shaping a new approach, one in which it is integrated within MAGTF operations.*

According to the Marines interviewed, the intelligence community views UAVs as "their assets" because that is how the system evolved in Afghanistan.

As one squadron member emphasized: *UAV operations personnel would basically check in with the air officer who would then pass them over to intel and they would then work together.*

Rather than having UAVs as part of the fire support system, they became assets which were part of observation and evaluation and the authorization of fires was handled separately.

As one squadron member underscored: *This became a loop rather than a straight line which is where we would like it be when we operate as a MAGTF.*

The separation of Marine Corps UAV assets was the norm rather than the exception. As one squadron member emphasized: *When I would fly in Afghanistan, I might look down and see a Shadow or Scan Eagle below me, but I never once coordinated with these assts. I had no idea what they were looking at. I just knew that they were below me,* noted the Operations Officer.

Lt General David A. Deptula, who in his last active duty position oversaw the planning, policy, and development of Air Force UAVs, and grew that force by over 500 percent in the Air Force, agreed with the Marine officers interviewed about the need for integration.

One of the biggest advantages of remotely piloted aircraft is that they allow for the condensation of the 'find, fix, and finish' kill chain onto one platform. To capitalize on this capability these aircraft need to be integrated into the entire combat enterprise, not just one piece of it.

That is exactly what the next phase of UAVs involve in the Marine Corps, the integration of these systems within the Air Combat Element (ACE) of the MAGTF.

As one squadron member noted: *The Ground Combat Element (GCE) should be requesting the capability, not the asset. If you need persistent IS with full motion video, that will probably fall to UAS.*

The UAS operator is a key part of the equation and when it works properly, the operator can work with the GCE and work with the sensor onto the target by shared situational awareness.

The challenge is shaping ways to parse the information to the appropriate element within the MAGTF to empower the GCE or ACE to become more effective.

A problem facing the USMC is the relatively limited capabilities of the UASs which they currently operate whether the Shadow or the RQ-21A. Their range is limited, and their footprint is not agile. The Marines are bringing the RQ-21A aboard amphibious ships but its limited range – 50 nautical miles – and its footprint limit its utility. And it takes up precious ship space as well.

As one squadron member emphasized: *4 Shadows take about 50-55 Marines to operate with only one airborne at any one time. With the RQ-21A we will take 22 marines with us onto the ship with the MEU but with our equipment size, we are already taking up about 12-15% of the space on the ship.*

These limitations also can frustrate the training and career processes. Marines are trained at Air Force UAV schools and the Air

Force personnel go on to operate Predators, and the Marines operate a much more limited asset.

Another clear requirement is to build swappable packages for the evolving USMC UAS birds as well, for missions can highlight C2, ISR or EW needs.

And the sweet spot for the USMC would be to have UASs, which could work with Ospreys and F-35s to provide persistent capability complementing the insertion force. The RQ-21A has too small of payload to be able to provide for this kind of operational flexibility being limited to the 20 pound or less category.

But there is a clear need, one driven by USMC innovation overall. As one squadron member highlighted: *When we looked at an after action report for a SP-MAGTF mission, there was a desire to have communications reachback, the ability to have armed escort and persistence surveillance, all capabilities which the proper UAS can provide. Why would put in anything else but a UAS to provide for those capabilities?*

And a core challenge facing UASs clearly is not just bandwidth but jamming. As one squadron member emphasized: *Our current platforms actually operate at bandwidths that are commercially available. Our Shadow operates at the same bandwidth as your WiFi does.*

Beyond the question of evolving a new generation of UASs more appropriate for the USMC expeditionary approach and Osprey and F-35 enabled and capable of collaborative engagement, there is an opportunity in the short term. The USAF has Predators going to the Air National Guard and as Afghanistan winds down Predators could be made available to the Marines.

For the Africa and Middle East missions, Marine Corps Predators could be operated on French bases in West Africa and Djibouti in East Africa (where the Italians recently operated Predators for several months) with the ground stations able to support USN-USMC ARG-MEUs operating on either side of Africa.

The Marines did not suggest this option, but they did focus on how longer-range land based assets could support an expeditionary force. As one squadron member expressed his judgement: *The Reaper is the best thing out there, whether you look at speed, range or endurance. And you can have the ground control station anywhere you want within the broad operational area.*

C2 for Hybrid War: The Marines Preparing for Combat

November 16, 2015

By Robbin Laird and Ed Timperlake

Counter-insurgency warfare has dominated U.S. military engagements for more than a decade, with joint warfare largely defined as air and naval services supporting ground forces doing COIN operations.

COIN has become so dominant that key elements of fighting forces have been crafted in its image: slow motion warfare, hierarchical C2, the growth of the OOLDA (Observe, Orient, Legally Review, Decide, and Act) loop replacing the decisive OODA loop, K-Mart type logistics capabilities, significant Forward Operating Bases, and uncontested airspace.

Under Marine Corps Commandant Amos, the shift from COIN dominance was recognized and initiated through the "return to the sea" or ramping up combat Marines' experience operating from the amphibious fleet. As Amos testified, *The Marine Corps is not designed to be a second land army"* but rather *"is designed to project power ashore from the sea.*[4]

The Bold Alligator exercise series launched in 2011 highlighted this return to sea-based operations. The maturing Osprey and arriving F-35B are powerful enablers for the Navy-Marine Corps team to shape expeditionary forces capable of inserting force, achieving objectives, and withdrawing across the range of military operations.

A key part of ensuring mission success is appropriate C2 to lead flexible insertion forces into and out of operations.

The Second Marine Air Wing recently held Wing Exercise 15 at Cherry Point to train for C2 flexibility supporting insertion forces where near-peer competitors might be involved. Col Kenneth Woodard, the exercise director, emphasized the challenge of wing-level training without higher headquarters present to simulate different command process players.

2nd Marine Aircraft Wing Marines initiate and assist with flight requests during Wing Exercise 15 at Marine Corps Air Station Cherry Point, N.C., Oct. 13, 2015. (Photo by Cpl. Unique B. Roberts).

The exercise simulated operating in land environments with near-peer competitors, requiring defensive and offensive actions to support forces and ensure operational success.

Woodard commented: *We had a near peer competitor with robust aviation elements and surface-to-air defenses to counter their offensive capabilities, and in our scenario, we were reacting to some of their attacks.*

Expeditionary logistics proved crucial for dynamic operations that cannot rely on pre-existing supply chains. The exercise established a FOB to support advancing forces, addressing critical questions:

- How do you supply it?
- Can you use trucks or KC-130s?
- Where can fuel be stored? How do aircraft access the FOB? How do you establish communications?

The exercise incorporated real-world complications like pilots getting sick and included sailors as medical team members simulating casualty management—crucial for leaving no one behind in today's combat environment where hostages can become political pawns.

Wing Exercise 15 serves as a prelude to exercises with 2nd Marine Expeditionary Force and Bold Alligator, with lessons learned folded into a dynamic learning process preparing Marines for hybrid warfare.

The Marine Corps is working to shape modern 21st-century insertion forces capable of operating across the full spectrum of military operations. This transformation requires moving beyond COIN's hierarchical, slow-motion approach to embrace the flexible, dynamic C2 necessary for hybrid warfare scenarios where multiple threat types from different adversaries create complex operational environments.

Success in this new paradigm demands expeditionary thinking, technological integration, and C2 structures that enable rapid decision-making and execution which is a fundamental shift from the bureaucratic overlay that characterized the COIN era.

ELEVEN

2019 Visits: Transitioning to Arctic Operations

The U.S. Marine Corps has been undergoing a significant strategic transition, shifting its focus from decades of Middle East desert operations to preparing for high-end conflicts in diverse environments, including the Arctic.

The USMC's participation in Trident Juncture 2018 demonstrated both capabilities and limitations. It showed they can show up effectively to support crisis management, but it remains necessary to shape the force integration in Norway to be able to gain the level of warfighting capabilities needed in a crisis.

This transition represents a significant strategic shift for the Marine Corps, requiring new training, equipment adaptations, and operational concepts. As global tensions continue to rise and the Arctic becomes increasingly contested, the ability to operate effectively in this harsh environment will be a critical capability for the USMC in the coming decades.

My visits in 2019 highlighted this transition.

2nd Marine Wing at Trident Juncture 2018: The Case of MAG-31

February 12, 2019

During my visit to Norway last year, I had the opportunity to visit several airbases and speak with Norwegian officers and defense officials about the shifting security landscape in Northern Europe.

With the return of direct defense challenges to the Nordics, there has been a major shift toward recapitalizing forces, introducing mobilization measures, and reworking concepts of operations to address the Russian threat. However, it has been nearly two decades since the Nordics faced a direct defense threat, and the nature of that threat has fundamentally changed.

Previously, the core threat was an amphibious assault from the Soviet Bloc similar to Germany's World War II operations. Now the threat is different, and the concepts of operations must adapt accordingly. As I wrote in *Front Line Defence*:

It is clearly not your daddy's Cold War but, for the younger generation, not having lived through it, it can be a bit of a shock facing a nuclear power that has threatened Northern Europe with destruction if they don't comply with Russian security demands.

But there is no Warsaw Pact. The Russians cannot lead an envelopment campaign against Northern Europe. In the Kola Peninsula, Russia maintains the greatest concentration of military power on earth, making Northern Europe a key flashpoint. The opening of the Arctic is changing the strategic geography as Putin establishes new military bases to provide greater reach into the North Atlantic.[1]

For the Nordics, Trident Juncture 2018 was a building block for addressing this new strategic situation. According to Col. Lars Lervik, a key Norwegian officer involved in exercise preparation: *"A key focus of the exercise from the NATO side is exercising our ability to conduct high intensity operations in a multi-national environment. We need to be able to function not only as individuals and individual nations, but actually function together."*[2]

The exercise also served as a command post exercise, providing Norway an opportunity to integrate multinational operational training into its defense modernization process. Col. Lervik emphasized: *It is very important to ensure that we have the procedures in place necessary to operate an integrated force on Norwegian territory in a higher intensity operational environment.*

An F/A-18D Hornet, with Marine All-Weather Fighter Attack Squadron (VMFA) 224, conducts an aerial refuel in Norway, during Exercise Trident Juncture 18, Oct. 24, 2018. Photo by Gunnery Sgt. Christopher Giannetti.

During my visit to 2nd Marine Air Wing, I spoke with Marines involved in the exercise to understand their perspective on Northern Europe's return to direct defense and the challenges facing coalition interoperability. I interviewed Col Pares, Commanding Officer of MAG-31, and LtCol Joshua Pieczonka, Commander of VMFA Squadron 224.

Both officers acknowledged that while the exercise was successful from a mobilization standpoint, years of Middle East operations had left them unprepared for cold weather operations. More critically, the exercise failed to achieve a core goal: training for the high-end fight. The Marines deployed and flew missions but did not effectively integrate with Marine Corps forces in Norway or with Norwegian and allied forces.

Col Pares explained: *We've been sending squadrons for years now to the sandbox in the Middle East, and it's been a long time since we sent anybody back to the North. Thanks to the Commandant's direction, we're attempting to get back into cold weather operating environments. We want to make sure our equipment is still able to operate there, and that we know how to operate there.*

LtCol Pieczonka noted that while the exercise succeeded at the tactical level, higher-end training capabilities needed improvement: *We did take eight F-18s, packed them up and moved them across the Atlantic Ocean. We were able to move essentially the majority of the VMFA and put it in place over a couple of days. From a strategic perspective, of getting something somewhere into a base and just figuring out operations and then having a very warm reception by our Norwegian counterparts, it was very well done.*

However, the exercise revealed significant gaps. There is a clear need to develop processes for Marines arriving in Norway during a crisis to bring their full combat capabilities and integrate effectively with Norwegian forces. This represents a work in progress that will require sustained attention.

The exercise highlighted technology integration challenges, particularly with the new G/ATOR radar system. While the Marines successfully transported the radar to Norway, they were unable to fully demonstrate its combat capabilities or achieve the connectivity and force integration necessary for effective operations.

Previous work at MAWTS-1 had focused on unlocking G/ATOR's capabilities for the USMC and joint force, including integrated fire control between G/ATOR and F-35 aircraft. The radar provides targeting information and fires support ashore, but realizing its full potential requires comprehensive integration with allied systems.

As noted in earlier MAWTS-1 evaluations: *We are going to be able to provide significantly greater information to all of the shooters, whether airborne, shipborne or ground based missile defense systems.* However, Trident Juncture 2018 demonstrated that this integration remains a work in progress.

The Nordic commitment to the F-35 presents an important opportunity to leverage emerging combat capabilities in the region and develop better competencies in command and control to optimize forces for high-end combat. Col Pares emphasized the

Commandant's significant focus on seeking cold weather training opportunities as the Marine Corps ramps up readiness and transitions from desert operations.

Trident Juncture 2018 represents an important turning point in this strategic shift. The Marines demonstrated they can deploy effectively to support crisis management, but substantial work remains to develop the force integration capabilities needed for warfighting in the Nordic environment.

The exercise revealed that while the fundamental mobility and deployment capabilities exist, achieving true coalition interoperability for high-intensity operations requires continued investment in training, technology integration, and procedural development. Future exercises must focus not just on getting forces to the fight, but on ensuring they can fight effectively together once they arrive.

2nd Marine Wing at Trident Juncture 2018: The Case of MAG-26

February 13, 2019

For the Nordics, the Trident Juncture 2018 exercise was a building block for shaping approaches to dealing with the new strategic situation.

During my visit to 2nd Marine Air Wing earlier this month, I had a chance to talk with Marines involved in the exercise to get their sense of the return of direct defense in Northern Europe and the challenges facing the Marines to provide the kind of force engagement which ultimately the Nordics, the U.S. and NATO would like to see in terms of coalition interoperability necessary to operate in a crisis situation.

I had a chance to talk with the CO of MAG-26, Col Boniface, whom I have met with several times before he took this command, and Lt.Col Fowler, the CO of VMM-365.

According to Col Boniface: *It is important to note that during the exercise, which encompassed actions in Iceland and Norway, the V-22 operated above the Arctic Circle. We were able to deploy, engage and provide presence in the exercise. We had to deal with the weather and operating conditions in the region, which are quite different from where our Marines have spent most of their time in the past decade. And we need to continue to learn how to operate in those conditions, and to have the domain knowledge of how to exercise patience and timing appropriate to operations in the Nordic region. The weather comes in, each fjord has its own weather so to speak and we have to learn patience and how to deal with the second and third order affects which operating in cold weather generates.*

Most of the conversation about the Trident Juncture 2018 engagement involving MAG-26 focused on the experiences of VMM-365 and LtCol Fowler provided an overview and various insights into the USMC experience.

According to LtCol Fowler, the impact of Hurricane Florence on North Carolina meant that they had reduced participation in the exercise. The initial plan was to send six aircraft, but they did send 114 Marines and 4 Ospreys to the exercise.

LtCol Fowler highlighted that they operated from the Iwo Jima amphibious ship and in Iceland did a raid against an "enemy" airfield. That raid was launched from the ship and the force returned to the ship after the raid. The raid did not highlight the long-range capability of the Osprey but rather operated as integral part of the insertion force which also included CH-53Es and related assets.

A major piece of the operations in both Iceland and Norway was working with the Osprey in cold weather conditions. Notably, they were operating the Osprey's de-icing capabilities and getting a comfort level with the aircraft in cold weather conditions.

LtCol Fowler underscored the point made by Col Boniface with regard to the importance of weather conditioning and learning in Norway during the exercise: *The Norwegians are great partners. They supported us as we worked our learning curve in the cold weather environment. But clearly we need to improve the communication systems used during the exercise, to get the full combat capability out of our force and to better integrate with the Norwegian force as well.*

And as all pilots note when flying in Norway, it is not just the weather, which is challenging but the terrain and the infrastructure built into the terrain as well.

A U.S. Marine Corps MV-22B Osprey with Marine Medium Tiltrotor Squadron (VMM) 365 conducts flight operations during Trident Juncture 18 at Vaernes Air Base, Norway, Nov. 1, 2018. Photo by Lance Cpl. Cody Ohira.

LtCol Fowler added: *With the towers and power lines running throughout the fjords, it is dangerous for aircraft operations. And we operate both as a helicopter and as an airplane so we faced challenges which are both the same but different for both type of craft all rolled up into one type of aircraft! There was extensive use of UASs as well during the exercise, which creates a challenge to sort out the operations of the manned with the unmanned aircraft operating in the same airspace as well. Clearly, this is a work in progress.*

One change which is critical to reshaping operations is the nature of the local community, meaning that when operating in Norway it was clear that they are a committed ally and the population was highly committed to supporting Marine Corps operations, including providing real time intelligence with regard to the "enemy" force. This was noted as a significant difference from USMC operations in the Middle East.

In short, the picture provided of MAG-26 involvement in Trident Juncture 2018 reinforced the picture provided by MAG-31. The exercise was a success in terms of being able to project force, but to get the full combat value from a Marine Corps force in a real crisis, significant effort needs to be directed towards enhanced capabilities to integrate the insertion force with the host nation and its force.

My discussions in Norway as well as Denmark have underscored how important shaping an effective C2 system for the defense of Northern Europe. In an interview with Brigadier General Rygg conducted last year at Bodø Airbase, the Chief of the Norwegian Air Operations Centre highlighted the importance of getting C2 right in the new strategic situation.[3]

Brigadier General Rygg: *We are building out new C2 capabilities within the National Joint Headquarters. It is about technology and reworking the workflow. We are bringing the key players into a close working relationship within the mountain to provide for better crisis management support as well.*

As infrastructure changes, the focus will as well to provide for crisis management support.

Brigadier General Rygg: *We are shifting from a classic joint targeting approach to a joint effects approach. Every time that you do something with the military, you are creating an effect. We are fielding new systems, which provide capabilities we have not had in the past. How do we use these systems to create the appropriate joint effect?*

The kind of C2 system needed is clearly an agile, scalable and flexible one. Brigadier General Rygg: *We may need to provide for mission control where the autonomy of key systems will be maximized. We may need to have a tight hierarchical C2 system. It depends on the threat; it depends on the mission and on the crisis management situation. But we need to build in redundancy and flexibility from the ground up.*

Clearly, the Marines agree and would underscore the core importance of enhanced interoperability in a crisis situation to get full benefit from working together.

The Osprey at 2nd Marine Air Wing: An Update from Colonel Boniface

February 14, 2019

It has been a couple of years since I last visited with the Osprey community at New River, and the discussion with Colonel Boniface provided some significant updates on their activities.

First, the Marines like the other services have been hit with significant readiness challenges rooted in the sequestration period. High tempo ops continued while support dollars were significantly reduced. This is hardly a recipe for success.

Col Boniface: *When I got here about a year and a half ago, the goal was to basically recover readiness. We've had every squadron has either deployed, deploying, or come home. And, with that being said, we also saw a significant dip in readiness. We've seen about an 85 percent increase in availability of aircraft over the last year and a half.*

Second, while the restoration of support dollars is clearly underway, the challenge is to put in place a more effective support approach to the Osprey fleet which is seeing new users put in place as well as the opportunity to put more effective global support in place as well. In other words, enhanced financial support clearly is a necessary but not sufficient condition to get the support enterprise right.

Col Boniface underscored: *One of the biggest challenges that I have here is mitigating the long term down process. My problems aren't necessarily operational. My problems are an insufficient supply system and a significant amount of corrosion that I have on some of these aircraft. The corrosion piece is being addressed but the corrosion piece takes awhile to get the engineers to come back and say that this aircraft is good. But, we don't have a suitable amount of the engineers and engineering support to be able to turn these corrosion problems and fix them quick enough and turn them into available aircraft.*

There is a significant opportunity to think through what the next round of logistical sustainment for the Osprey fleet could be. As Col Boniface put it: *We need better predictability, forecasting, and availability of parts. It's very difficult to identify where your next supply shortfall is coming from and that's where this community struggles.*

I added the comment that somebody needs to be thinking through the re-crafting the sustainment enterprise so that the money is put into non-repeatable mistakes and ensures a more predictable and sustainment logistics support enterprise.

There clearly is a need for a healthy supply system built on supporting global operational realities and this challenge will become even more significant with the strategic shift in operations underway.

Col Boniface added: *We just need a better supply model which can level out the supply chain support to the deployed force.*

Third, when I first visited New River several years ago and talked with the Osprey training squadron, there major focus was upon Marines and the Air Force. Now with the U.S. Navy buying Ospreys as well as the Japanese, there are new stakeholders in the training process, and that training squadron has become a priority effort within MAG-26 for sure.

Fourth, the Osprey is hitting its mid-term life cycle and will need upgrades, which will enable the aircraft to continue being effective going ahead. And along with upgrades, the challenge of repairs associated with corrosion, a normal challenge for a sea-borne fleet needs to be addressed as well.

Fifth, MAG-26 like other elements of the USMC are facing the challenge of shifting from the Middle East land wars, as a primary focus, to work in regions quite different against peer competitors.

The specific case we discussed was the engagement of MAG-29 in Trident Juncture 2018, where really for the first time for Marines who had operated the Osprey for many years in the Middle East, they had to deal with Cold Weather and the dynamics of weather in Iceland and in Norway.

According to Col Boniface: *It is important to note that during the exercise, which encompassed actions in Iceland and Norway, the V-22 operated above the Arctic Circle. We were able to deploy, engage and provide presence in the exercise. We had to deal with the weather and operating conditions in the region, which are quite different from where our Marines have spent most of their time in the past decade.*

And we need to continue to learn how to operate in those conditions, and to have the domain knowledge of how to exercise patience and timing appropriate to operations in the Nordic region. The weather comes in, each fjord has its own weather so to speak and we have to learn patience and how to deal with the second and third order affects which operating in cold weather generates.

In short, MAG-26 is in good hands but face significant challenges as the force is reworked to deal with the new strategic environment, and one in which a more effective logistical enterprise needs to be put in place for global operations.

As the Navy and the Japanese join the Osprey nation perhaps this will be easier to get done.

The USMC and a New Chapter in Heavy Lift: The CH-53K Logs Demo at New River

January 29, 2019

A decade ago, the author witnessed the challenging transition from the CH-46 to the Osprey, where maintainers grappled with a completely different aircraft that marked aviation's shift into the digital age. Now, the Marines face another transformational moment with the CH-53K, which despite looking similar to its predecessor CH-53E, represents an even more dramatic leap from mechanical to digital systems.

The CH-53K may appear to be a cousin of the CH-53E, but it's fundamentally different. The aircraft has been designed from the ground up to simplify operations by significantly reducing mechanical parts and integrating health maintenance systems directly into the aircraft configuration rather than as bolt-on additions. Learning from the Osprey experience, Marines are taking a proactive approach to ensure comprehensive support for the CH-53K fleet from day one.

Unlike previous aircraft transitions where logistics were treated as an aftermarket consideration, the Marines are integrating overall logistics sustainment infrastructure into their evolving concepts of operations. This holistic approach represents a revolutionary change for the USMC that could easily be overlooked without understanding the strategic significance of what they call the "log demo."

At New River, an impressive team of Marines, NAVAIR officials, and Sikorsky field representatives are working on the log demo using an aircraft that will become part of the first operational squadron. The effort is led by VMX-1 detachment personnel, including LtCol Jade Campbell, an experienced CH-53E operator with recent Australian Ministry of Defence experience, and LtCol Stu Howell, an experienced CH-53K pilot with Presidential helicopter program background.

The transformation extends to how pilots will operate the aircraft. LtCol Campbell explains that while CH-53E pilots developed core competence in hovering under difficult conditions, the CH-53K's digital systems handle these challenges automatically. As Campbell notes, *Unlike with the CH-53E, there's no challenge hovering a Kilo on a moonless night in a dusty zone. And that allows the CH-53K pilot to have the mental bandwidth to think about the battlespace.*

This represents a generational shift from pilots primarily focused on operating mechanical aircraft to those who can concentrate on battlespace awareness and their role within broader mission objectives while the aircraft handles functions previously requiring pilot attention.

The log demo encompasses several critical activities. The baseline work involves taking aircraft maintenance manuals and testing every procedure to determine what works, what needs modification, and what new procedures might be more effective. As LtCol Howell explains, *We're not redesigning the aircraft, but we are creating better procedures or mitigating fixes to help improve the safety of the aircraft.*

Beyond procedure verification, the team is examining how skill levels and workforce composition must change for CH-53K operations. They're developing new organizational structures for heavy lift maintenance departments, questioning traditional approaches like whether 18-year-old high school graduates with one year of training can handle the sophisticated maintenance requirements of this digital aircraft.

The focus extends to supporting the "fly, fix, fly" operational tempo that the Kilo will require, ensuring logistics infrastructure enables rather than impedes operations. The team looks beyond individual aircraft to fleet management, working with NAVAIR and Sikorsky at Pax River to leverage big data generated by the fleet. The goal is positioning the supply chain to empower operations by ensuring the right parts are available when needed, rather than following rigid, arbitrary delivery schedules.

The U.S. Navy is addressing broader support enterprise questions, including optimizing parts movement for shipboard operations and ensuring supply chains can support both land and sea-based CH-53K operations. As LtCol Howell notes, even seemingly simple logistics like transporting rotor heads to Connecticut for overhaul require careful consideration of available resources at bases like New River and Cherry Point.

The lessons learned from the log demo are being directly transferred to the initial operating squadron co-located at New River. The next phase involves taking four aircraft through initial tests and evaluation, which will become the foundation for the first operational squadron and training instructors.

This comprehensive approach to integrating the CH-53K represents more than just introducing a new aircraft. It's reshaping how the Marines think about heavy lift operations in the digital age. By treating logistics, maintenance, and operations as integrated elements rather than separate concerns, the Marines are positioning themselves for a more effective and sustainable heavy lift capability that truly represents 21st century air systems revolution.

Preparing for Effective Fleet Support: The CH-53K Log Demo at New River

January 29, 2019

With the coming of software upgradeable aircraft which have health management systems built in, there is a strategic opportunity or better put, a strategic imperative, to leverage those systems to gain knowledge and mastery of a combat fleet.

The opportunity is there to gain predictive knowledge about fleet performance and to shape a workforce and approach to enhanced aircraft availability and better ability to deal with the required ops tempo.

A case in point is the CH-53K. It is a digital rich aircraft with its health maintenance system built in. And the Marines along with NAVAIR and Sikorsky are currently working the logistics side of the aircraft at New River to determine best procedures to maintain the aircraft, and how to structure the workforce and shape a logistics infrastructure, which can be optimized for fleet operations and support.

Recently, I visited the Marine Corps-NAVAIR-Sikorsky team working the logs demo at VMX-1 to get an overview on the approach.

An issue facing software upgradeable aircraft such as the K is the concurrency challenge. This challenge has most frequently been identified with the F-35 but it is at the heart of the change which software upgradeability brings to a fleet.

Because software drops can be placed directly into combat aircraft much more rapidly than the historical cycle of modernization, the new capabilities need to be reflected in both the pilot and maintenance simulators as well to ensure that there is integration of training, and operations across the fleet.

But the gap, which occurs between the software drop in the aircraft and the software on the simulators for both maintainers and pilots, is the concurrency challenge and one, which clearly needs to be addressed.

During my visit, I had a chance to talk with Jim Lambert, the head field representative for Sikorsky working on the Log Demo. He is a very experienced CH-53 Marine who has worked with Sikorsky for a number of years in support of CH-53s operating worldwide. He has brought that operational fleet experience with the aircraft from before its birth and is being deployed to work with the CH-53K through its migration from factory to the logs demo to the first operational squadron.

Obviously, that kind of domain knowledge is crucial to getting the most effective combat aircraft to the force.

A key aspect of what we discussed was the opportunity to build out fleet knowledge from which aircraft availability would be enhanced over the experience of flying earlier generation CH-53s.

The photo is credited to VMX-1 inn 2019 and shows Marines working on the rotor head which is a major piece of what allows the aircraft to carry three times the weight externally compared to a CH-53E. The elastomeric Rotor Head is designed to be much lower maintenance compared with the Rotor Head on the earlier version of the heavy lift helicopter and allows the blades to be removed and remounted significantly faster compared to the CH-53 E.

According to Lambert: *We're actually collecting fleet-wide diagnostic data which has not been done before on this scale. Every aircraft is contributing to a metric showing usage and trending all these data points.*

Once the data is collected it is analyzed to identify any negative trends and adjust as needed to optimize the fleet and increase aircraft availability. This allows the fleet to identify possible part issues or new failure modes at the earliest possible point. This allows the operator to validate its logistics foot print real time and to be predictive with fleet needs to put parts in the system ahead of need.

This is all done in the background of the user ensuring the maintainer has what they need when they need it. This sets up for a proactive based approach to fleet support as opposed to the current reactive approach.

Lambert argued that working as a team from the ground up and having field representatives intimately familiar with the maintenance of the aircraft was crucial to getting the right kind of support in operations that the CH-53K would need.

As Lambert put it: *As field reps, we will stay on this flight line as long as we need to keep training Marines and assisting them troubleshooting airplanes. Our goal also is to bring fleet continuity of experience to ensure that the Marine maintainers are comfortable with the airplane and not doing what you have described as the Mongo challenge or corrective.*

We're here to say, "No, it really is okay what you are doing. And here's why it's okay."

Our main goal is to remove insecurities generated by lack of familiarity or operating from the way the CH-53E is maintained and to try to provide them a little bit of comfort with the new approaches to sustaining a new aircraft. Because we're in this crawl, walk, run environment as we stand up the CH-53K

In short, the Marines are pioneering a 21st century approach to maintaining a 21st century software upgradeable aircraft. This is clearly not a CH-53E but a whole new animal which requires a whole new approach.

Working the Logistics Con-Ops as the CH-53K Enters the Force

August 31, 2019

The heavy lift transition from CH-53E to CH-53K represents one of the final modernization efforts of Marine Corps Aviation. This essential capability replacement for the readiness-challenged 30-year-old CH-53E leverages lessons learned from previous transitions of the MV-22, H-1, and F-35 aircraft.

The shift from CH-53E to CH-53K is significant for the USMC, representing a transformation from a mechanical to a digital aircraft. The CH-53K has been designed from the ground up with sustainability in mind, with Marine maintainers involved from the outset to facilitate more rapid and effective logistical support.

At New River MCAS, the first CH-53K has arrived for the Marines. Marine Corps logisticians who participated in the aircraft's design are now joined by other logisticians working through maintenance procedures before the aircraft joins the first operational squadron. These "loggies" are shaping the template that will be refined and evolved by the operational units.

The logistics demonstration team, led by LtCol Stu Howell and including SSgt Curtis A. Kelly, SSgt Jeremy C. Lombard, and SSgt Mike V. Farina, is working through maintenance and logistics procedures. This team combines Sikorsky representatives and Marines, with many Sikorsky employees being former Marines.

The team's core finding: Everything is much more straightforward to work on. The design process has yielded an aircraft that's easier to maintain in terms of system accessibility, while the digital nature eliminates several systems present on the CH-53E. The flight controls represent a key improvement due to the fly-by-wire system.

As the team noted: *This aircraft is much more plug and play compared to the CH-53E. A lot of the systems on the CH-53E have been eliminated with how the CH-53K has been designed and built.*

The transition involves three major shifts:

- Design Simplification: The design-for-maintainability process has simplified operations.
- Digital Transformation: The shift from mechanical to digital systems removes maintenance-intensive mechanical component.
- Validation Challenge: Working with computer information and sensor systems requires validating and refining maintenance procedures.

The team emphasized that *log demo is synonymous with maintenance evaluation and validation; we are not flying the aircraft. We are assessing the procedures and improving them.*

Access to dynamic components is significantly easier compared to the CH-53E. The new aircraft eliminates "miles of cable and boxes" in favor of a "install a box, validate and move forward" approach.

A critical aspect of the log demo involves marrying maintenance procedures with appropriate tools. Since Marines operate in both shipboard and austere environments, ensuring the match between procedures and required tools is essential.

Key considerations include:

- Do we have the right torque wrench?
- Do we have the right socket?
- Do we have the right stand to work on gear box?

This extends to engine boxes and containers necessary for transporting maintenance tools and parts. The focus remains on sustainable deployability and expeditionary operations, with the goal of running "out of the gate" rather than slowing initial operations due to incomplete procedures.

Similar to the F-35 transition, maintainers must understand the aircraft as a complete system rather than working as federated maintainers focused on individual components. There's a need to understand how individual expertise contributes to overall aircraft performance.

The team stressed: *We need to revisit training to understand how the magic boxes work overall with regard to the whole aircraft. Everybody needs to be digitally competent. The component specialists need to understand how the aircraft talks to us digitally from a systems point of view.*

This aircraft represents a clear departure from being "CH-53E 2." The transition involves a shift to conditions-based maintenance, which is a significant change that makes navigating the maintenance process easier through digital information systems.

The team concluded: *It is not that hard to move from the CH-53E to the CH-53K. But not sure that those who will be CH-53K guys could make a transition to working on a CH-53E. You are more of a technician on the CH-53K; with regard to the CH-53E, you are more of a mechanic.*

This transformation reflects the Marine Corps' evolution toward more sophisticated, digitally-enabled aviation maintenance capabilities that will enhance readiness for future Marine Air-Ground Task Force operations.

Visiting the Warlords: An April 2019 Update from the CO of VMFAT-501

April 29, 2019

During my most recent visit to 2nd Marine Air Wing, I had a chance to visit MCAS Beaufort and meet with LtCol Adam Levine, the CO of VMFAT-501, otherwise known as the Warlords.

As the base was busy for the airshow being held the weekend of the 26th of April, the CO graciously provided some time for an update on the USMC training efforts and shaping the pipeline for the training aspect of the fast jet transition in the Marine Corps.

I first dealt with the Marines getting ready for F-35 as Eglin stood up the first training efforts. But I have not been back to Beaufort for four years, and the Marines have been busy ramping up their training efforts during that period.

LtCol Levine provided a comprehensive update on those efforts. It was obvious from the flight line that more planes, pilots and maintainers were populating the base since I was last there.

The CO highlighted that as the operational squadrons gained experience in executing the various missions in which the aircraft is involved that operational learning was being brought back to the training effort and providing greater accuracy with regard to the demand side but also the training effort was able to work better training for preparing for operational missions.

The command has obviously scaled up since the last time I was there with more than 100 pilots trained and with the standing up of the second training squadron at Beaufort over the next few months, that scaling up would be accelerated as well.

The challenge is a significant one as the USMC will transition from their legacy force to an all F-35 one within the next two decades and the task of the training squadrons will be to train the "newbies" and the experienced pilots from legacy aircraft to fly and operate the F-35.

The training cycle is eight months during which the pilots learn to fly the new jet and then to take the jet through its paces with regard to variety of missions for which the Marines use their fast jets.

When I was last there, no "newbies" were present; only experienced pilots. Now the "newbies" are the majority of pilot trainees.

I asked the CO who is an experienced Hornet pilot how the two cohorts experience was different. It must be remembered that heart of fifth gen aviation is a man-machine revolution, where the pilot is getting comfortable with the performance of his aircraft generating data providing situational awareness and the pilot interacting with his screens while operating the aircraft.

He made the point that the "newbies" had never experienced the much more pilot intensive processing of data which legacy pilots do, expected their machines to work in ways that could facilitate what they wanted to do, but to do them faster.

In other words, they already assumed the new baseline of man-machine interaction and wanted that interaction to speed up. The pilots of legacy pilots had much more appreciation of the fact that the F-35 was working from a very different baseline than their legacy jets did.

The training of the two cohorts was handled a bit differently as the more experienced combat pilots could do more training in the simulators with the "newbies" doing more time in the cockpit.

I wanted to discuss with the CO the challenge of training with regard to a software upgradeable aircraft. I have discussed this challenge with regard to other software upgradeable aircraft, in Williamtown Airbase with the RAAF and the P-8 with Jax Navy.

Put simply, the advantage of the software upgradeable aircraft is that the historical type/model/series understanding of an aircraft now transitions the type by the software enabled combat systems on board and which variant is onboard the particular aircraft or squadron of aircraft.

This is the concurrency issue, which is built into a software upgrade process, although the defense press has incorrectly only identified this challenge with that of the F-35.

My photo taken during the 2019 Beaufort Air Show.

Not a surprise because they IOCd first and to the operational impacts from operating these aircraft in the Pacific and are transitioning their initial 2B software jets to 3F and this transition requires both a hardware and software upgrade.

What this means is that pilots in the training process need to become familiar with both variants of the aircraft and understand the interaction of the two. This is not a bad thing because in the operational world they will need to work with aircraft operating globally which are at various software levels, both with regard to services and partners.

LtCol Levine has been flying the F-35 for more than seven years and has witnessed first-hand the software roadmap taking shape from block 1A through 3F. He underscored that the evolution onboard the Hornets flown by the Marines compared to flying the early variants of the F-35 did not demonstrate the generational differences which now are evident with the 3F.

According to Levine: *There is simply no comparison between a 3F F-35 and a legacy aircraft. They are in different worlds.*

The Marines at Beaufort have and are working closely with allies. The Brits stood up their training at Beaufort and have jets, pilots and maintainers working with the squadron until this summer. Now the training squadron is being stood up at RAF Marham, and the RAF and Royal Navy will train there.

But with the departure of the Brits, the Italians are coming next and will train for the next couple of years before their carrier comes to the U.S. for final certifications in a couple of years.

TWELVE

2020 Visit: Working the Integrated Distributed Insertion Force

In the face of evolving global challenges, the United States Marine Corps is undergoing a significant transformation, shifting from counterinsurgency operations in the Middle East to a focus on distributed operations and naval integration in what is often called the "great power" competition.

This evolution is driven by the need to operate effectively in contested environments against peer competitors while maintaining the Marines' traditional expeditionary capabilities.

Exercise Deep Water: Working the Integrated Distributed Insertion Force

December 31, 2020

In July 2020, North Carolina-based Marines conducted Exercise Deep Water, a significant training event that demonstrated the evolving approach of the Marine Corps to distributed operations and integrated force insertion.

According to II Marine Expeditionary Force's November 5, 2020 press release, *Exercise Deep Water involved Marines from 2nd Marine Division, 2nd Marine Logistics Group, and 2nd Marine Aircraft Wing* conducting operations at Marine Corps Base Camp Lejeune. The exercise featured two battalions conducting an air assault to command and control various capabilities organic to II MEF in preparation for major combat operations.

The 2nd Marine Division described Deep Water as nearly double the size of the previous year's Exercise Steel Pike, making it the largest exercise of its type conducted on Camp Lejeune in decades. The exercise was planned and led by 2nd Marine Regiment, incorporating elements from across the II Marine Expeditionary Force Marine Air Ground Task Force (MAGTF).

Exercise Deep Water featured a dynamic force-on-force scenario against a "peer-level adversary," simulated by 2nd Marine Division's Adversary Force Company. As Colonel Brian P. Coyne, commanding officer of 2nd Marine Regiment, explained, "Exercise Deep Water, a regimental air assault that utilizes the whole of CLNC and the outlying training areas, will allow us to sharpen our spear and help make us more lethal."

The exercise incorporated multiple phases of operations:

- Air assault operations with robust aviation support.
- Offensive, defensive, and stability operations.
- Multiple urban training scenarios.
- Operations against both conventional and hybrid adversary forces.

The training built upon 2nd Marine Division's priority to enhance readiness against peer threats, aligning with both the National Defense Strategy and the Commandant's Planning Guidance.[1]

During a December 2020 interview, Major Rew, the exercise's air mission commander, provided detailed insights into the technical aspects of Deep Water. The exercise combined forces from pickup zones in North Carolina and Virginia, executing a force insertion into a contested environment.

The operation utilized a comprehensive array of aircraft:

- USMC Assets: AH-1Z, UH-1Y, F/A-18A/C/D, AV-8B, MV-22B, CH-53E, KC-130J, and RQ-21.
- USAF Assets: F-15E and JSTARS.

The exercise sequence involved air assets first clearing areas for the Assault Force, establishing air superiority, then inserting the assault force via MV-22B Ospreys and CH-53E helicopters. The Marines moved significant numbers of personnel from two locations hundreds of miles apart to nine different landing zones.

A key innovation was the use of an Osprey operating as an airborne command post, equipped with a "roll-on/roll-off" C2 suite providing chat capability and interoperability with ground, mobile, and static command posts. Major Rew emphasized the advantage: *Having the NOTM-A kit on the Osprey is a big win because it provides so much situational awareness. With the Osprey as a C2 aircraft, there is added flexibility to land the aircraft close to whatever operational area the commander requires.*

The exercise leveraged MAGTF Tablets (MAGTAB) for digital interoperability, enabling real-time information sharing between ground elements and aviators. The MAGTAB provided visual representation of integrated effects and outcomes to the command element. ISR support came from both USMC assets and a USAF JSTARS aircraft, utilizing the Network-On-The-Move Airborne (NOTM-A) system to provide interoperability for commanders and assault forces.

Major Rew highlighted the future integration of F-35 aircraft into air assaults and distributed operations: *They're an incredible sensor and they have the capability to see what's happening on the battlefield, assess things real time, and then send that information to the individual who needs to make a decision. Incorporating them into future exercises of this magnitude will be value-added to the entire Marine Corps.*

The exercise utilized the Networking on the Move (NOTM) Family of Systems, a SATCOM-based command and control capability initially fielded in 2013. NOTM is an ACAT IV(M) program with a $509 million budget across the Future Years Defense Plan and $1.7 billion total life cycle cost.

NOTM provides:

- Robust C2 wideband SATCOM capability.
- Three external network enclaves (SIPR, NIPR, and Coalition).

- Access to Global Information Grid and Next Generation Enterprise Network.
- Full motion video, VoIP, and VoSIP capabilities.
- Integration with ruggedized laptops and tactical software.

U.S. Marines with 2d Marine Air Wing take off after dropping Marines with 2d Battalion, 2d Marine Regiment, 2d Marine Division as part of Exercise Deep Water on Camp Lejeune, North Carolina, July 29, 2020. (U.S. Marine Corps photo by Lance Cpl Brian Bolin Jr.).

Exercise Deep Water represents the Marines development of an ecosystem for integrated and distributed force insertion. As new ISR, C2, and strike capabilities enter the force, they can be integrated into this evolving ecosystem. The exercise demonstrated the Marine Corps' commitment to maintaining readiness for major combat operations while adapting to peer-level threats.

Colonel Coyne concluded, *This training event will improve our warfighting proficiency and prepare us for tomorrow's battles. 'Tarawa' Marines will fight and win if called.*

The exercise showcased the Marine Corps' evolution toward distributed operations, integrated command and control, and multi-domain warfare capabilities essential for future conflicts against peer adversaries.

Visiting MAG-29 at 2nd Marine Air Wing: The Perspective of Col Finneran

December 23, 2020

During a visit to 2nd Marine Air Wing in early December, I met with Col Robert Finneran, Commanding Officer of MAG-29, to discuss the future of assault operations. Col Finneran brings extensive combat experience from Iraq and Afghanistan, along with tours at MAWTS-1 and SOCOM, primarily flying AH-1W and AH-1Z aircraft. As CO of MAG-29, he now serves as the Naval Aviation Enterprise's fleet lead for the CH-53, including the transition to the CH-53K.

When asked why the CH-53K's differences from the CH-53E are difficult to convey, Col Finneran emphasized the dramatic technological advancement: *It starts with the silhouette of the two aircraft. They are very similar, but that is about it. The Kilo [CH-53K] is a generational leap in technology. It is a completely different airplane as far as capabilities and technology.*

The challenge lies in the aircraft being Marine Corps-only, limiting joint force recognition of its revolutionary capabilities. The CH-53K's digital systems significantly enhance situational awareness, while its improved speed and range provide substantial advantages for force insertion.

Anything we can do to enhance the situational awareness for the pilots, and take the workload off of them, allows them to focus on their mission, Finneran noted. While CH-53E pilots will find flying the new aircraft relatively easy, the CH-53K generation must learn to manage the wealth of information available to maximize mission utility.

The aircraft's aerial refueling capability transforms risk calculations for force insertion. *We can extend range and move a force and needed sustainment quickly and across great distances. This provides both the Marine Corps and the Joint Force Commander flexibility and complicates the problem for any adversary.*

2020 Visit: Working the Integrated Distributed Insertion Force • 187

Col Robert B. Finneran, left, converses with MajGen Karsten S. Heckl, center, and SgtMaj Jacob M. Reiff, right, at Marine Corps Air Station New River, North Carolina, March 30, 2020. (U.S. Marine Corps photo by Cpl. Paige Stade).

Col Finneran highlighted the significance of digital interoperability coming to the H-1 family, particularly for Marine Corps integration with the U.S. Navy. Early discussions with Seahawk Weapons Schools focus on complementary employment of airframes.

We bring a lot to the table for maritime operations, notably in terms of the weapons we carry on the Viper, he explained. Once networked into the maritime kill chain, these aircraft offer substantial capabilities for both sea control and sea denial operations.

A critical discussion centered on mobile, expeditionary basing and the role of logistical connectors. The CH-53 family serves as a crucial enabler, capable of delivering fuel directly to expeditionary bases without requiring multiple staging locations.

Fuel is certainly critical to distributed operations and our heavy lift helicopter is a key enabler, Col Finneran emphasized. *I only see it increasing in importance to such operations. I don't see how the force goes and does any of the new operational concepts without that capability.*

Col Finneran argued that his force is ideally suited for

distributed operations. The attack utility team created by the Viper and Venom, combined with CH-53 heavy lift capability, provides rapid maneuverability for desired combat effects compared to slower basing methods. This approach offers sustainability by bringing mission-essential materials rather than depending solely on ship-based support.

The CH-53K's performance improvements in range and payload provide "a lot more options" compared to the CH-53E for expeditionary scenarios.

MAG-29 collaborates with MAWTS-1 on developing new Tactics, Techniques, and Procedures (TTPs) to maximize capabilities for evolving basing requirements and combat effects needed for sea control and denial operations.

Col Finneran concluded with a strategic perspective: *If we really want to be risk worthy, and we really want to challenge the risk calculus for our adversary, we've got to outmaneuver them in both time and space, and that is what my command is focused on delivering to the fight.*

The visit underscored three key themes: the CH-53K represents a fundamental technological advancement that will reshape Marine Corps capabilities; digital interoperability will enhance naval integration; and Marine Corps aircraft serve as essential integrative elements ensuring distributed forces remain connected rather than isolated and dispersed.

Visiting HMLA-269 and 167: Shaping a Way Ahead for Marine Corps Light Attack Helicopters

December 30, 2020

Marine Light Attack Helicopter Squadron 269 (HMLA-269) is currently commanded by LtCol Short and is a squadron consisting of the AH-1Z Viper attack helicopter and the UH-1Y Venom utility helicopter. They are known as the Gunrunners, and next door within the same hangar are the Warriors of Marine Light Attack Helicopter Squadron 167 (HMLA-167), commanded by Lt. Col. Hemming. They are part of Marine Aircraft Group 29 within the 2nd Marine Aircraft Wing.

The "Gunrunners" operate what they refer to as "attack utility teams." What that means is that they operate the Viper (AH-1Z) and the Venom (UH-1Y) as an insertion and support package. They share 80% commonality of parts, operate from a small logistical footprint and are extremely maintainable in the field which make them a significant expeditionary warfare asset.

Recently, both squadrons retired their last AH-1W Super Cobra attack helicopter in favor of the AH-1Z Viper. As LtCol Hemming noted in the interview: *The hydraulics, the engines, and some of the systems on the aircraft and the air frame, are significantly more durable and reliable than the old AH-1 Whiskey in terms of the amount of hours you can put on before you have to conduct maintenance on it. These significant upgrades result in your ability to operate the aircraft for an extended periods of time compared to the legacy Cobra and Huey.*

The Viper brings significant firepower to an expeditionary unit with the Venom providing lift and support to that unit as well. The helicopters have evolved from their legacy ancestors to be more capable as well.

As LtCol Short put it: *We are the most expeditionary and resilient attack helicopter platform there is in terms of the scale and the ability to survive in the field or operate forward. Our hydraulics, our control systems, our powertrain systems are the most expeditionary maintainable as an attack utility team in operation today. We are the, as somebody described it, 'The punchy little friend in the overhead that's there when no one else is".*

In the counter-insurgency environment, the attack utility team could operate in a distributed environment to support Marines fighting toe-to-toe against insurgents. LtCol Short argued that their attack utility team was very "risk worthy" in terms of *the logistical, the manpower, the cost investment for the capability gain, you would give a ground force, or you would give a supported force by putting them forward, putting them into a position to offer support.*

The Viper is adding Link-16 and full motion video so that it can be even more supportable for or supported by an integratable insertion force.

It is also very capable because of its relatively small footprint able to land in a variety of ground or ship settings and get refueled.

If one focuses on the ability to operate virtually in any expeditionary setting, at sea or on land, the Viper is extremely capable of refuelability for an insertion force. They can do this onboard virtually any fleet asset at sea or at a Forward Air Refueling Point or FARP.

From a concept of operations perspective, notably with regard to an ability to operate from multiple bases, the attack utility package certainly can keep pace with the "pacing threats" facing the Marines.

The Commandant has asked the Marines to rethink how to do expeditionary operations, and to promote tactical innovations to do so. HMLA-269 has been focused on this effort.

Notably, they have been exercising with the Ground Combat Element at Camp Lejeune to work small packages of force able to be inserted into the combat space and able to operate in austere locations for a few days to get the desired combat effect and then move with the GCE to new locations rapidly.

HMLA-269 has been working closely with 3rd Battalion, 6th Marines to shape innovative ways to deploy expeditionary force packages. LtCol Short noted: *We are working ways to work distributed force operations with the battalion.*

They have a security mission currently with regard to II MEF in reinforcing Norway. The question being worked is: how, in a multi-basing environment, can one provide the kind of firepower that the maneuver force would need?

The Gunrunners took a section of aircraft to work with a ground combat unit and to live together in the field for a period of time and sort out how best to operate as an integrated force package. They operated in the field without a prepared operating base and worked through the challenges of doing so. They worked with an unmanned aircraft ISR feed as part of the approach. Obviously, this is a work in progress, but the strategic direction is clear.

And there are various ways to enhance the capability of the force to be masked as well. Movement of small force packages, operating for a limited period of time, moving and using various masking technologies can allow the attack utility team which is operational now to be a key player in shaping a way ahead for Marine Corps expeditionary operations.

In short, the attack/utility team of 2nd MAW are taking the force they have, and their significant operational experience and adapting to the new way ahead with the next phase of change for expeditionary warfare.

An Update on the CH-53K from VMX-1: The Perspective of LtCol Frank

January 5, 2021

During my visit to 2nd Marine Air Wing during the first week of December 2020, I had a chance to visit New River Marine Corps Air Station and meet with LtCol Frank, VMX-1, to get an update on the coming of the CH-53K. LtCol Frank showed me the simulator as well giving me a chance to experience the flying qualities and, notably, the ability to hover via using the automated systems to operate in difficult visual and operating conditions.

He joined the USMC in 2002 and has flown a wide variety of rotorcraft during his career and served as a pilot for the U.S. President under President Obama. He came to VMX-1 in 2018. He has stayed in large part to follow through the CH-53K to fruition, that is into operations.

As LtCol Frank put it: *It is crucial to have a CH-53 fleet that works effectively as it is a unique capability in the USMC crucial for our way ahead operationally. It is the only aircraft we have that can move an expeditionary brigade off of our amphibious ships.*

We have about a hundred Marines here at the test detachment. We've been training our maintainers and our air crew on the 53K for two years now. The maintainers have been working on it since 2018, when we started the logistics demonstration which is essentially the validation of maintenance procedures on the CH-53K. I have 10 pilots in the det including myself and I'm responsible for ensuring that everyone goes through the proper training syllabus.

All 10 of our pilots in addition to our crew chiefs and our maintainers will be the first unit to be allowed to operate a "safe aircraft for flight" which is a term we use for the maintainers.

Our job is to conduct initial operational test and evaluation training for six months, beginning this month and ending in May or June of 2021, where we will establish five aircraft commanders, myself being one of them, five co-pilots, that'll be our 10 pilots. We'll qualify 10 crew chiefs, and our maintainers will continue to advance in their maintenance quals. In June of 2021 is when we enter into IOC evaluations.

We're going to evaluate the reliability and maintainability of the aircraft. We're going to collect all our maintenance data, determine how long it takes to fix, how long it's down before it's fixed and how many flight hours it accomplishes per maintenance man hour to evaluate it.

We will evaluate its shipboard compatibility in June and July 2021. We are to evaluate its desert mountainous capabilities in Twentynine Palms, beginning of August and September 2021. And we also have a sorties generation rate demonstration where we will execute a surge capability of sorties from a ship in November 2021; we'll do that for a period of about 72 hours straight, where we will fly every aircraft every day and see what they deliver.

We discussed the importance of the fly by wire system in the aircraft, which he considers "very mature." He did note that the USMC subjects its aircraft to some of the harshest environments in the DoD, salt spray, open ocean, desert heat and freezing cold. Robustness is a crucial aspect of determining reliability.

We do not operate runway to runway. We do not store them inside; we use them in challenging conditions.

He referred to his team as "the learning curve for the CH-53K," similar to what happened with the Osprey or the F-35B.

He underscored that the aircraft is well along the path to IOC. *We've had a lot of time with the aircraft. Our Marines have been working on it for two years now. During logistics demonstration, we took the publications, which were in their infancy, and we went through every work package. The bulk of the Marine Corps' CH-53K personnel, equipment, aircraft, and support will be located at VMX-1 when the Marine Corps declares the CH-53K program is IOC.*

LtCol Frank described the innovation cycle as follows: *When problems come up with the aircraft, we bring up to the program office, the program office sends it out to engineering and industry. They implement changes. They implement engineering fixes, and they incorporate them.*

While at New River, we visited the first of the CH-53Ks delivered to VMX-1, which I had seen earlier in the log demo program but now was on the tarmac.

LtCol Frank indicated that VMX-1 is to receive six aircraft overall. *We are to receive our next aircraft on January, February, June and September of 2021, and the last one on January of 2022. By January 22, when the sixth aircraft is delivered, we should be done with IOT and E and we should carve out a detachment size group of maintainers, pilots, and aircraft from VMX-1 to form the initial cadre of HMH-461.*

How does he compare the CH-53Es to the CH-53Ks? *I've started in the Ch-53D in 2004, they're my first love. I'll always love them. They were much harder to fly. And the ease of flying this, the flight control system is probably the biggest game changer for the 53 community.*

We're not used to anything like this. It's very intuitive. It can be as hands off as you know, a brand-new Tesla, you can close your eyes, set the autopilot and fly across country. Obviously, you wouldn't do that in a tactical environment, but it does reduce your workload, reduces your stress.

And in precision hover areas, whether it's night under low light conditions, under NVGs, in the confines of a tight landing zone, we have the ability to hit position hold in the 53 K and have the aircraft maintain pretty much within one foot of its intended hover point, one foot forward, lateral and AFT, and then one foot of vertical elevation change.

It will maintain that hover until the end of the time if required. that's very, very stress relieving for us when landing in degraded visual environments. Our goal at VMX-1 is to create tactics that employ that system effectively.

Some communities struggle with how they use the automation, do they let the automation do everything? Do they let the pilots do everything? How to work the balance?

We're working on a hybrid where the pilots can most effectively leverage automation.

If you know you're coming into a brownout situation or degraded visual environment, you engage the automation at a point right before the dust envelops you. And then in the 53-K, you can continue flying with the automation engaged.

You continue flying with the automation engaged, and you can override it, but as soon as you stop moving the controls, it will take your inputs, estimate what you wanted and keep the aircraft in its position.

It's a very intuitive flight control system, and it blends very well with the pilot and the computers. It allows you to override the computer. And then the second that you stop overriding it, the computer takes back over without any further pilot input. That's probably the biggest game changer for our community.

Visiting MAG-26: The Perspective of Col Spaid

January 4, 2021

During a visit to Marine Aircraft Group 26 at New River Marine Corps Air Station, I met with Col John Spaid, a veteran Osprey pilot who has been instrumental in the aircraft's development since 2005. His experience provides unique insight into the MV-22B Osprey's introduction, evolution, and operational impact on Marine Corps aviation.

Col Spaid was part of the first Osprey deployment to Iraq in 2007 under LtCol Rock (now MajGen Rock) and participated in the first MEU deployment in 2009. The introduction strategy involved three groups of Marines operating six-month rotations over 18 months, with Spaid's squadron being the first to enter Iraq flying from the Middle East.

Having previously flown CH-46 helicopters with HMM-263 in Iraq, Spaid immediately recognized the Osprey's revolutionary capabilities. *We could now operate at altitude with speed and range and able to circumnavigate the battlespace and then insert in a favorable point,* he noted. The aircraft's survivability was enhanced through new airframe materials, providing tactical advantages unavailable to legacy aircraft.

The transition from conventional helicopters to the Osprey presented significant challenges. The fly-by-wire capability and unique flight control system were entirely new to the rotary-wing community. Spaid initially doubted his ability to achieve precision landings, joking that he thought "it was going to be an area weapon" during simulator training. However, he quickly mastered the aircraft's capabilities.

The tactical advantages became clear in combat operations. *Being able to operate high and fast while minimizing our time in the climb and dive then coming in unexpected, that was a tactical advantage that no one else had or seen,* Spaid explained. The operational difference was dramatic; missions that required weekly CH-46 flights from al-Assad to al-Qa'im became daily operations with the Osprey.

The early Osprey community developed a strong esprit de corps, calling themselves the "Osprey Nation." Despite small deployment numbers, the team brought diverse aviation backgrounds that informed their operational approach. *It was a melting pot of Marine Corps aviation,* Spaid recalled. *We all brought our best professional military aviator qualities into this effort, which means we had a unique opportunity to filter out bad habits that may have been lingering in our previous communities.*

The community focused heavily on developing tactics, techniques, and procedures (TTPs) from combat experience. Lessons learned from dust landings and reduced visibility operations in Iraq were brought back to inform follow-on squadrons. From 2010-2012, standardization became crucial as West Coast squadrons began standing up.

In 2009, Spaid deployed as Aviation Maintenance Officer with the first MEU Osprey deployment under squadron commander Col Paul "Pup" Ryan. Operating in the CENTCOM area with Fifth Fleet, the composite squadron included 29 aircraft, with 12 Ospreys split between an LHD (eight aircraft) and LPD (four aircraft).

This deployment required the Navy to adapt to Osprey operations while Marines learned shipboard maintenance procedures. *We could fix it faster, we could launch it faster, we could fold it faster than originally expected,* Spaid noted, building Navy confidence in the platform.

The deployment's success influenced combatant commanders, leading to the development of Special Purpose MAGTFs and what became the North Africa Response Force (NARF). As Spaid observed, *We could offload in Kuwait but operate all throughout Iraq. That was an eye opener for them.*

Reflecting on his role as a "plank holder" in Osprey Nation, Spaid emphasized that the historical significance wasn't apparent at the time. *We were just doing our jobs. You focus on mission accomplishment; you don't really understand the historical significance of the event at the time.*

The transformation has been remarkable. From four aircraft on the New River tarmac in 2007 to today's substantial fleet, the change reflects both the aircraft's maturation and the Marine Corps' evolution. *All our Marines are smarter*, Spaid noted. *Baseline pilot intellect is now through the roof. We've normalized what we thought was creative thinking and training.*

As commanding officer of MAG-26, Spaid's primary focus is "sustainment, sustainment, sustainment." The aircraft continues receiving software upgrades targeting reliability and maintenance processes, with close collaboration between the program office and industry partners to maintain aircraft reliability.

The Navy's decision to acquire the Osprey for resupply missions brings multiple advantages: enhanced sustainment prioritization across Naval Aviation Enterprise, shared training opportunities at New River, improved parts availability across fleets, and collaborative upgrade opportunities. The integration promises enhanced Navy-Marine Corps cooperation, particularly valuable given new emphasis on joint operations.

International partnerships have also proven successful, particularly with Japanese pilots and maintainers training at New River. *As new FMS opportunities arise, we'll be able to leverage off the great success we have had with our Japanese allies*, Spaid noted.

Col Spaid's journey from the Osprey's combat introduction to leading MAG-26 represents the aircraft's evolution from experimental platform to proven capability. His experience illustrates how the "Osprey Nation" transformed Marine Corps aviation through innovation, standardization, and relentless focus on mission accomplishment. As the platform continues evolving with Navy integration and international partnerships, the foundation built by pioneers like Spaid ensures continued success for this revolutionary aircraft.

Multiple Basing, Kill Webs and C2: Shaping a Way Ahead

January 7, 2021

During my visit to 2nd Marine Aircraft Wing in early December

2020, I had a chance to discuss these challenges with the C2 professionals in Marine Air Control Group 28.

We had a wide-ranging conversation on the intersection between the evolving tactical environment and C2 and will highlight a number of takeaways from that conversation. To be clear, these are my conclusions shaped by the discussion, but I am not attributing these conclusions to the group.

As the United States and core allies develop an integrated distributed force, a critical challenge is shaping command and control (C2) systems that fully enable such operations. With sensor networks increasingly integrable with C2 systems and the focus on distributed forces, the key question becomes: how do we connect distributed forces with joint or allied elements that provide critical combat mass during crises?

The kill web approach recognizes that combat clusters must carry their own C2 capabilities to integrate effectively in operations. Reach-back to larger forces depends on networks — both intelligence, surveillance, and reconnaissance (ISR) and C2 — that may be denied during conflict. Force packages need built-in integratability, with broader reach to other elements shaping how they can affect the wider battlespace.

Given challenges from core adversaries, flexible basing has gained importance. Sea basing represents a core advantage, combined with multi-domain forces operating from various bases that can intersect with sea bases. For Marines, this means reworking amphibious forces alongside new basing approaches.

Based on discussions with Marine Air Control Group 28 professionals during my visit, five major challenges emerge:

- Crisis Management and Integration: How will different crisis situations be managed with distributed forces? The focus on operations short of total war emphasizes escalation management and control. Key questions include how Marines can assist in sea denial, sea control, and controlling sea lines of communication while

ensuring proper integration for crisis management or deterrence effects.
- Transitional Challenges: Moving from current capabilities to required future states presents immediate obstacles. The amphibious force lacks the Navy's most advanced C2 systems, and insertion operations aren't built around sea-base management of shore forces. This requires more robust, expeditionary C2 aboard amphibious ships and new capabilities connecting expeditionary forces ashore with those afloat.
- Avoiding Combat Orphans: Forces at expeditionary bases need adequate C2 capabilities to achieve missions within integrated operations. This involves building webs of C2 nodes or "node basing." Distributed C2 remains challenging, with rapidly deployable systems more aspiration than reality. Integrating amphibious task forces must precede pushing small combat teams ashore with current technology limitations.
- Decision-Making Authority: Operating from various bases raises questions about tactical decision-making authority, particularly regarding fires solutions. Mission command alone insufficient when shaping integrated fires with mixed launch points, including expeditionary bases.
- C2 and ISR Acquisition Reset: Historical platform dominance over seamless C2 and ISR data flows requires fundamental change. How will the necessary C2 capabilities be built into forces to achieve required integratability for distributed operations?

A crucial question emerges about applying expeditionary basing across the Atlantic versus Pacific theaters. Russia poses the most direct threat to the United States through Putin's actions, nuclear buildup, and Kola Peninsula force projections threatening U.S. forces and territory directly. This reality drove 2nd Fleet's 2015 reestablishment and deepening U.S.-NATO relationships along the UK-Nordic-Polish defense arc.

This creates opportunities to reshape the relationship between 2nd Marine Aircraft Wing (MAW) and 2nd Fleet. An arc from North Carolina through Norfolk, Halifax, Newfoundland, Iceland, Greenland, to Norway and the Nordics offers deployment opportunities for Marine force packages emphasizing strike, anti-submarine warfare, anti-surface warfare, C2, and ISR support missions.

Leveraging 2nd MAW's air capabilities would be crucial for this reshaping function. Lessons learned could then apply to Pacific operations, potentially more significant than traditional island-hopping approaches.

Effective C2 integration remains crucial for shaping survivable, sustainable, and effective distributed forces within overall maritime campaigns. The solution requires extensive training and exercises to determine optimal distributed force structures.

As one discussion participant emphasized: *We need to increase our C2 communication dynamics in our exercises. We need to exercise our vulnerabilities and to find ways to enhance our strengths.*

This conclusion aligns with the core focus on training for high-end conflict. Admiral Nimitz's World War II directive of "training, training, training" translates to today's need for "exercises, exercises, exercises" to develop more effective distributed forces capable of operating in contested environments.

The challenge ahead involves not just technological solutions but fundamental rethinking of how distributed forces integrate, communicate, and maintain effectiveness across multiple domains and geographic regions while preserving the flexibility and survivability that modern threats demand.

THIRTEEN

2021 Visit: New Operational Concepts

The United States Marine Corps finds itself at a pivotal moment of transformation, adapting to new strategic challenges while leveraging technological advances in aviation, simulation, and operational concepts.

Based on a series of interviews conducted with Marine leadership in 2021, I was able to examine how the 2nd Marine Aircraft Wing (2nd MAW) was training and engaging in the transition of U.S. forces to deal with the evolving strategic environment.

Advancing the Future Force with Integrated and Realistic Simulation: A Visit with the Leadership of 2d Marine Air Wing's Training Systems Leadership

September 12, 2021

During a July 2021 visit to 2d Marine Aircraft Wing (MAW), I talked with LtCol J. Eric Grunke and LtCol Jessica Hawkins. Grunke just relinquished responsibility to Hawkins for the Marine Aviation Training Systems (MATS) ecosystem within 2d MAW, encompassing all aviation simulators and necessary infrastructure across MCAS Cherry Point, MCAS New River, and MCAS Beaufort.

Maj J. Eric Grunke, pictured here at Marine Corps Air Station Cherry Point, N.C., April 24, 2012 has been named Marine Corps Aviator of the Year by the Marine Corps Aviation Association. The MCAA gives the award to the pilot who makes the most outstanding contribution to Marine aviation over that past year. Credit Photo: USMC

In order to remain America's Force in Readiness, the Marine Corps must continue to adapt as the world changes. Major General Michael Cederholm, Commanding General 2d MAW, emphatically believes that training must evolve exponentially to strengthen Marines' enduring advantages and allow them to prevail in strategic competition with China or any other nation.

China has fundamentally transformed the operating environment, and the Marine Corps must modernize the force and its capabilities to continue deterring adversaries. The CG directed the integration of multiple, disparate platform simulators along with command and control (C2) systems to better prepare his force and support II Marine Expeditionary Force missions.

A key element of the combat learning process is integrating live, virtual, constructive training — a technique that combines simulation with real-world flights and ground maneuver. This technique is a force multiplier when shaping tactics and concepts for new and emerging technology like the F-35.

Advances in modeling, simulation, and workforce integration

provide an alternative to traditional war gaming. Commanders that leverage advancing capabilities in virtual and constructive environments are provided with a dynamic operational environment that truly exercises real-time risk and force employment decision-making at all levels. More importantly, it allows commanders to engage a "thinking" enemy and the associated friction often lacking in static assumptions used by traditional war gamers.

LtCol Grunke's experience in Operation ODYSSEY DAWN provides an example of operational plans not playing out to script. While flying Strike Coordination and Reconnaissance sorties, he had to quickly change mindset and serve as the on-scene commander for an innovative downed pilot recovery effort using a combined flight of Ospreys and Harriers. His actions were a potential war-changing event that war game modeling can easily overlook.

Much of my discussion with LtCol Grunke and LtCol Hawkins focused on how to enhance integrated training and increase pilot proficiency against enemy aircraft and weapons systems. A critical step is having real pilots operating in their platform-specific simulators and integrating with the Marine Aircraft Control Group (MACG) to conduct coordinated missions.

This concept was recently tested in 2d MAW's COPE JAVELIN exercise. As highlighted in an April 29, 2021 article by 1stLt Michael Curtis, the exercise followed a fictional operational scenario that could easily take place in the real world. Marine aviators from various units across 2nd Marine Aircraft Wing strapped into flight simulators for different aircraft located at different bases across eastern North Carolina, connecting across different simulation systems to work together defending against a fictional enemy force.

This integration gives Marine pilots and Marine Air Control Group 28 Marines the opportunity to accomplish hard, realistic training without leaving their respective bases and saves tremendous amounts of money in fuel, ordnance, maintenance, and various other costs associated with conducting this training in real time.

Achieving this level of integration was described by LtCol Grunke as a crawl, walk, then run process:

- Crawl phase: Get a Cobra pilot in their sim and a Harrier pilot in their sim to share the same visual representation, allowing them to fly and see the same terrain while working together in that common operating picture.
- Walk phase: Take that pairing and work with a Joint Terminal Aircraft Controller in their sim to execute a single sortie. In this phase, the MACG (the C2 arm of the wing) was purely constructive, with personnel traveling to various sites to participate but not on their own equipment.
- Run phase: Seen in COPE JAVELIN, the C2 element at the DASC controlling close air support assets coming in and out of objective areas on their own gear, while Harrier pilots in their sims and Cobra pilots in their sims participate in the engagement.

Several themes emerged from our discussion:

- Networking Integration: The challenge of linking individual platform simulators to deliver a more integrated operating space remains significant. As LtCol Hawkins explained, all systems have been developed somewhat independently and speak their own language. To work around this, all networking information must run through a Distributed Information System (DIS) bridge, which interprets various coding languages used by each simulation device and converts it into usable language for each to understand.
- Visual Database Consistency: Training in the environment where operational plans are intended to be executed requires comprehensive visual databases. Currently, different platforms have purchased visual imagery databases based on their own assessed priorities, resulting in disparities between platforms. For example, a

MV-22 might be able to go to Northeastern Europe in the simulator but the Harrier may not.
- Allied Integration: The allied aspect of training for integrated combat effects presents unique challenges. Norwegian F-35As carry a much different ordnance load from the F-35B flown by the USMC and have different rules of engagement considerations, requiring constructive role players from Norway to indicate how they approach operations.

As LtCol Grunke emphasized, *Instead of war gaming, let's train for a real war. Let's get into the areas where we expect to fight, with the actual terrain in a simulator, with the G/ATOR where we think it's going to be, where we think the force is going to launch from, and see how we do.*

When asked what key capabilities would most accelerate the training way ahead, LtCol Hawkins identified two priorities: building and operating from a common global visual database, and simulators that are easily connected to one another and can transfer data back and forth with ease.

The goal is clear: shape readiness at the wing level through working mission rehearsal that achieves true force-on-force exercises, where Marines can see whether their tactics work given their assumptions as scenarios unfold. This represents a fundamental shift from traditional training to a more dynamic, integrated approach that better prepares Marines for the complex challenges they may face in future conflicts.

LtCol Frank: A July 2021 CH-53K Update

September 4, 2021

During my July 2021 visit to 2nd Marine Air Wing, I had a chance to visit the VMX-1 CH-53K detachment at New River Marine Corps Air Station and to continue my discussions with LtCol Frank, Officer in Charge of the CH-53K Operational Test Detachment at New River.

The author with Lt. Col. Franks at New River, July 13, 2021.

Lt. Col. Frank provided an update on progress through the testing process but we took the opportunity to discuss as well the wider impacts which the CH-53K has on training and on operations as the USMC works its evolving approach to crisis management as part of the high-end fight.

Since my visit in December, LtCol Frank indicated that they had received new aircraft and had begun and then ramped up the flying hours. With their flight certification, they have now flown around 235 flight hours on the aircraft. They have certified five aircraft commanders, five co-pilots, 10 crew chiefs and more than a required number of maintainers with the appropriate level of qualifications for the next phase of training. That next phase will occur in August at 29 Palms.

They have completed their initial operational training but are waiting for certification to begin initial operational test and evaluation. In the meantime, they have engaged in a number of "rehearsal test and evaluation" sessions with Marines at 2nd MAW and Camp Lejeune to prepare for the August training efforts at 29 Palms.

The digital aircraft has many advantages and one can be seen on the training dimension. As with the F-35, pilots can train to core proficiencies more rapidly, which leaves room for expanding training options for the evolving mission sets which the Marines are clearly focusing on for full spectrum crisis management.

With regard to conversion training, they have discovered at VMX-1, that hours and flight events could be reduced for the pilots. As LtCol Frank put it: *The initial conversion syllabus from the CH-53E to the CH-53K was tailored based on our best guess of what events and flight hours would be required for the conversion aircrew. Following our initial foray into our own flight and simulator training and through our evaluations of the current syllabus we realized we could reduce those numbers by around 25%.*

Currently, we are focusing heavily on the co-pilot series-conversion syllabus which began as 17 total flight events for 26 flight hours. After our pilots completed this period of instruction, surveys taken at the end indicated that we could pare those numbers down by 7 events and 10 less flight hours. My hope is that this 25% savings will result in a typical Marine Heavy Helicopter Squadron saving 6 months over the duration of their transition.

So now, if we can can capitalize on the flight hours savings and pair that with an enhanced focus on the higher-level syllabus, we could expand training for those missions to meet high end events that the Marine Corps has decided is important in the evolving context.

We then discussed what he saw as the clear advantages of the Kilo over the Echo for the USMC. As he put it: *There is nothing sexy about assault support. Horsepower's our weapons system, and reliability is the key to providing the horsepower for the heavy lift needed for assault support.*

Reliability is crucial; and the CH-53K is focused on enhancing reliability over the legacy aircraft. As he put it: *If the grunts want a lift and they need six to eight helicopters, it would take a whole MAG effort of 53 ECHO to put the package in the air for a battalion. With more reliability, we would not need a whole MAG to do this.*

We're hoping that's where the CH-53K is going to help, with its digital systems, engine, rotor, and drivetrain system reliability. The Full Authority Digital Electronic Control (FADEC) provides enhanced control, health monitoring, maximum power and efficiency as often as possible. They also provide what we call automatic power assurance checks and integrated power assurance checks.

So we know exactly how engines are performing all the time. And it's providing real-time data.

Automatic means the FADEC is just pulling numbers all the time. It's a behind the scenes process. It's just going all the time and it gets downloaded onto our maintenance data card, which then the maintainers will plug into their ground module, their ground computer, and they can see the engine health.

Also, we can initiate power assurance checks as pilot, and the pilot can then bring up the summary of those and I can see, okay, power is doing good. Based on the spec engine performance, I'm actually plus 38 from the spec engine. So I've got more power than even the spec engine should have. The engine power available and limitations will be reflected on the primary flight display so we can be aware of that in the plane.

Such accuracy and certainty is critical when you do a high altitude and a high ambient air temperature lift. That's when the Kilo would be power limited. Knowing exactly how much power the engines are putting out, if I'm called to extract a platoon of Marines from a mountain top that's very high and it's very hot, and I have a lot of fuel on board, so I might be power critical, I can do a power assurance check and know that I'll be able to do it. Unlike the Echo, the CH-53K will give you a visual readout of your power status in real time while you are executing the lift.

LtCol Frank then addressed the reliability piece which the Integrated Vehicle Health Monitoring System (IVHMS) delivers.

Our main gearbox pressure sensor will say it's starting to fail or it's getting a false reading. It's still performing, but it's getting a false reading. And what our maintenance Marines will do is they'll interpret that maintenance data when we give them the data card and they'll say, "Okay, your main gearbox pressure sensor reported itself. Your intermediate gearbox reported itself for vibrations. That means there's a bearing failing in it.

As opposed to the ECHO where we would fly, and we would see chip light, caution light, oil pressure failures in the gearbox. That means the gearbox literally seizes or fails itself. That's when we know it's failed. In the CH-53K we'll get proceeding indications of that. Ideally, it leads to parts being removed before they fail. That should lead to increased maintenance readiness.

Things fail a lot in the legacy aircraft. As a flight crew, you build an anecdotal seat of the pants data base. I have had dozens of hydraulic system failures, *multiple engine failures, oil system failure, electrical components failing, attitude gyros failing at night and in IMC.*

All those things create the seat of the pants sense that you need a lot of hours to accumulate, those failures help you get the experience you need.

One benefit of these machine-aided pilot systems in the CH-53K clearly is that the less experienced pilots can approach capability levels of more experienced pilots. This will enable the man-machine system to deliver more safety for flights, and enhanced combat capability for the Marines as well. Assuming you're an experienced pilot, you have combat experience from which you could make judgments. But if I'm a less experienced pilot, now I have actually some machine aids that can help me.

Given that Marines are onboard one is talking about a lot of lives. And when the USMC Commandant and the Sergeant Major of the Marine Corps visited VMX-1 at New River in March, this was a key point which LtCol Frank underscored. Pilot vertigo can be a Marine killer and has been in past accidents. With the ability to push a button and let the aircraft fly itself, this should not happen in the future.

The advantages of a digital aircraft are very clear. But this means as well that cyber threats need to be dealt with on an ongoing basis, and clearly, the CH-53K program is not only aware of this but working it. Regular upgrading of software on the aircraft is part of the solution as well as cyber defensive capabilities as well. Both are being pursued with regard to the aircraft and its support systems.

What does LtCol Frank conclude with regard to the aircraft coming to the Marine Corps?

With the CH-53-K I would fly it 1,000 times over with my hair on fire before I would set foot in an ECHO again. Don't get me wrong, I love the old iron, still wear a CH-53D patch. I cherish my time in that plane. It's my first love. It's like an old Jeep, simple and reliable but unrefined, the ECHO is similar.

However, most of the time I would prefer to drive the Denali, that is the KILO. It's operational capabilities are much enhanced over those legacy aircraft simply by the awareness and aides it can provide to the flight crew; our crew chiefs and maintainers feel the same.

Being a generational shift, the new digital aircraft is in LtCol Frank's words "a blank slate."

You have an aircraft that can carry significant supplies or Marines inside and can carry 36,000 pounds externally. They can carry a lot of stuff. It has automated flight control systems that allows you to land in the degraded visual environments that you would not dare land an ECHO or a DELTA in. It can fly long distance without the air crew being fatigued. If you're aerial refueling and flying 1,000 miles in the CH-53E, the air crew would be wet noodles getting out after the flight. In the CH-53K you can relax a little, take a breath, allow the aircraft to help you fly and thus reduce aircrew fatigue significantly.

I think when the necessity for conflict rears its head the Kilo will be able to respond, and using human ingenuity, the operators will be able to find a way to support any mission that the Marine Corps needs it to do. The CH-53 K is so versatile that I don't see people being pigeonholed into not being able to do something with a CH-53K. I think they'll be able to answer the call 99.9% of the time.

It'll be able to pick it up. It'll be able to transport it, fly it any distance and land it anywhere. And you're not going to be afraid to do it. In the ECHO, if it was low light at night, the visibility was bad, you didn't have a moving map, and you were headed to a dusty and tight zone the pucker factor would be through the roof. The altitude hold was suspect, it didn't have lateral navigation and flight director capability, your attitude gyros would fail often. So you get this hair on the back of your neck stands up. I don't want to be flying in this environment. The aircraft's not going to help me, and I can't help myself because I don't have my sensory cues.

But in the CH-53K, you know the aircraft's going to help you. We've sat in brown out dust, just sitting there hovering and talking to each other with position hold on. And we've been debriefing the landing, and the aircraft's just holding a hover perfectly.

So that's what I like about the CH-53K is that I think it will be able to answer the call for the mission most anytime the Marine Corps needs it, whether we know what the mission is going to be, or not.

Shaping a Way Ahead for the TACAIR Community: Visiting MAG-14

August 11, 2021

During my July 2021 visit to 2nd Marine Air Wing, I had a chance to meet with the leadership of MAG-14.

I met with Col Williams, the MAG-14 Commanding Officer, LtCol Harrell, the MAG-14 Executive Officer and Maj Cunningham, the MAG-14 Operations Officer. Based at Cherry Point, MAG-14 operates AV-8B Harriers, KC-130Js and the RQ-21A Blackjack. It is in transition from the Harriers to F-35s, and their KC-130Js are key enablers for the entire MAGTF.

2nd MAW includes Marine Fighter Attack Training Squadron 501 (VMFAT-501), the Warlords, which is an F-35 training squadron at MCAS Beaufort.

But the coming of operational F-35s to MAG-14 and 2nd MAW will be another driver of transformation of 2nd MAW capabilities. Operationally, 2nd MAW deploys all over the globe, to include recently working with allied F-35s in the North Atlantic and European theaters in the recent BALTOPS-50 exercise.

But transition is always challenging, and one can see significant construction in preparation for the standup of the F-35 at the base. As Col Williams put it: *We are in transition as we wind down the Harriers and get ready for the arrival of the F-35s in the 2023 timeframe. We are scheduled to receive the first six aircraft in late FY23, while VMU-2 will transition from the RQ-21 to Reapers in the FY25 timeframe. We will continue to support the East Coast MEUs with Harriers, and will be sustaining the Harrier force through fiscal year 2028. When the F-35s take over the East Coast MEU's duty that will represent a significant transformation.*

The first hangar is being built as well as the simulator building for the F-35s coming to the base. As Col Williams noted: *We will eventually have two more hangars. A new headquarters will be built for MAG-14 as well.*

The challenge is to make the transition, but to maintain the readiness of the current force. Managing the two dynamics is the challenge which the MAG-14 leadership is facing. The Harriers deploy with the MEUs and as the MEUs transition towards more North Atlantic defense missions as opposed to Eastern Mediterranean missions.

The pilot and maintainer force will transition as the F-35s come onboard at Cherry Point as well. And this aspect is a key one in managing the transition as well; the Service has used incentive programs, such as Selective Retention Bonus to encourage reenlistment and continued service of highly qualified Marines, particularly aviation maintainers.

We then discussed the KC-130J. My own observation over the years is that the KC-130J is crucial to Marine Corps operations by providing logistic support, air-to-air refueling and close air support to fleet operating forces. As a multi-sensor image reconnaissance and close air support platform, the KC-130J aircraft may be equipped with the Harvest Hercules Airborne Weapons Kit (HAWK) configuration as well.

The Marine Corps has integrated the Harvest HAWK system, that provides the Battlefield Commander with a limited, persistent surveillance capability with the onboard Production Target Sight Sensor (TSS). The TSS can also provide the ability to employ precision weapons using laser guidance.

A core focus for MAG-14 is ensuring the readiness of the aircraft and crews for the KC-130Js. Given the aging inventory of aircraft, this is a key challenge going forward.

There has been a new focus on the long-range firing function which the USMC could participate in as they address evolving concepts of operations for extended littoral operations, such as the Marine Littoral Regiment. The MLR is a purpose built unit designed to enable the Marine Corps' new service strategy and employs three uniquely designed subordinate elements: a Littoral Combat Team, a Littoral Anti-Air Battalion, and a Littoral Logistics Battalion.

The LCT is designed to provide the basis for employing multiple platoon-reinforced-sized expeditionary advance base sites that can host and enable a variety of missions such as long-range anti-ship fires, forward arming and refueling of aircraft, intelligence, surveillance, and reconnaissance of key maritime terrain, and air-defense and early warning.

It has seemed to me that the Harvest HAWK experience could be leveraged here in terms of either working with longer range missiles, or adapting a Harvest HAWK capability for the Ospreys to provide rapid insertion fires into the fight. There is also the clear possibility that airlifters can be modified by using missiles in the box to be able to carry weapons that can be launched from the back of the aircraft.

Clearly, kill web approaches can allow for that, and setting up advanced expeditionary bases of C2 or sensor operating Marines supporting air-delivered fires is more expeditious than trying to move first to EABOs themselves.

MAG-14 is in transition, but it can be viewed as maintaining the competitive edge within a larger transition of the USMC itself.

VMFA-115 Trains with the Finnish Air Force

August 6, 2021

During my visit to 2nd Marine Aircraft Wing in July 2021, I spoke with MAG-31 and the VMFA-115 Operations Officer about their squadron's training with the Finnish Air Force.

Originally, VMFA-115 was to participate in Arctic Challenge 2021, a multi-national exercise. However, COVID-19 restrictions transformed their engagement into a bilateral exercise with the Finnish Air Force called ILVES. This provided an important window on how one might modify training going forward.

What VMFA-115 learned was how the Finns fight, how they operate their air force in a truly distributed manner, use their roads for landing sites, provide distributed logistical support, and work under the shadow of Russian long-range fires. Marines learning to fight as the Finns fight is valuable for understanding concepts like distributed maritime operations, littoral operations in a contested environment, and expeditionary advanced base operations—all crucial for operating from the High North to the Baltic Sea.

U.S. Marines with Marine Fighter Attack Squadron (VMFA) 115 fly alongside Finnish Air Force Fighter Squadron 31 over Rissala Air Base, near Kuopio, Finland, June 18, 2021. Marines with VMFA-115 are deployed to Kuopio, Finland in support of Squadron Visit ILVES (Finnish for "Lynx"). The purpose of Squadron Visit ILVES is to conduct air-to-air operations. (U.S. Marine Corps photo by Lance Cpl. Caleb Stelter).

Major Simmermon, VMFA-115 Operations Officer, described the experience: *A year ago, we were preparing for Arctic Challenge 21. If we had participated in that exercise, it would have been a big mission planning exercise and very scripted. We would have most likely used our own tactics and tried to incorporate what other countries were doing.*

But it became a bilateral exercise called ILVES. We were able to train with them in their tactics. A great tactic VMFA-115 was able to observe was the Finns diverting and spreading out to reduce the effects from a potential strike on their location. They showed us how they're able to set up expeditionary arresting gear, where they put their support and how they taxi the aircraft.

We then had one of their instructor pilots get in their simulator with us, where we practiced road landings — just taking off and landing on small, short expeditionary runways. The whole system relies largely on the logistics support and the infrastructure for their road runways, which are already in place.

Doing the bilateral training during ILVES exposed us to smaller level tactics, techniques, and procedures, which I had never seen before. Those conversations and briefs would not have been available in a big exercise like Arctic Challenge. It reminds you that even as a globally deployable force, it's important to see that there are different ways and different geographical locations that change the way an aviation unit fights. Seeing how other nations fight was very valuable.

When I visited Finland in 2018, I discussed with Lt. General Kim Jäämeri, former head of the Finnish Air Force, the unique way the Finns use their air combat capabilities. As he put it: *It is becoming clear to our partners that you cannot run air operations in a legacy manner under the threat of missile barrages of long-range weapons. The legacy approach to operating from air bases just won't work in these conditions. For many of our partners, this is a revelation; for us it has been a fact of life for a long time.*

The Nordics have ramped up cooperation in airpower integration through cross-border training. From 2015 on, the Norwegian, Swedish, and Finnish air forces have shaped a regular training approach that is flexible and driven at the wing and squadron level. They meet each November to set the schedule for the next year, but execution remains very flexible with a bottom-up approach.

The impact on Sweden and Finland has been significant in terms of learning NATO standards and enhancing capability to cooperate with NATO air forces. The training range being used is very large and uncluttered, allowing for great training opportunities and multi-domain operational training.

Since 2018, the Marines have ramped up their efforts to train in the Nordic region and operate in cold weather. With the Nordics increasing their defense capabilities and working greater integration with each other and North Atlantic partners, there are enhanced opportunities for Marines to work in the region.

VADM Lewis, who led the re-establishment of 2nd Fleet during this period of increased Nordic defense cooperation, emphasized learning from regional partners: *That has been my mantra from day one here: learning from our regional operations. As we work how best to operate in the region, we are learning from our regional partners some of the best ways to do so.*

For the Carolina-based Marines, this means expanded opportunities to learn from Nordic partners as they work enhanced integration with the U.S. Navy. The VMFA-115 experience with the Finnish Air Force demonstrates the value of this approach to cross-learning and adaptation for modern distributed operations.

Shaping a Way Ahead for the Assault Support Community: Visiting MAG-26

August 3, 2021

I first visited Marine Aircraft Group 26 (MAG-26) in 2007 when they were beginning the MV-22B Osprey transition. Now, the Osprey is the backbone of the Marine Corps combat assault support community. With both heavy-lift and light-attack helicopter squadrons, the Second Marine Aircraft Wing (2D MAW) is the cornerstone of all rotorcraft support for North Carolina-based Marines.

During my recent visit, I discussed the way ahead for combat assault support with three MAG-26 members: Maj Mazzola (MAG-26 Operations Officer), Maj Kevin O'Malley (VMM-263), and Maj Tom Gruber (VMM-365). We focused on the challenges of transition from the Middle East land wars and shaping a new way ahead for the assault force.

Up front, the shift was described by one participant as *A total paradigm change.*

2nd MAW Marines will perhaps lessen expeditionary operations in Mesopotamia and increase engagements in the North Atlantic area operations. Since 2018, focus on specific challenges such as cold weather training and exercises have increased at a pace not witnessed since the Cold War.

Case in point, one interviewee noted that in March 2022, they would once again train with the Norwegians in the Cold Response — one of the largest Norwegian and Coalition exercises since the Cold War. About 40,000 soldiers will participate in Norway's Cold Response 2022 exercise, planned to take place in the Ofoten area with the country's navy and air force as the main players.

General Eirik Kristoffersen, head of the Norwegian Armed Forces, noted: *There is a significantly increased interest among our allies for the north and the Arctic. It will be the largest military exercise inside the Arctic Circle in Norway since the 1980s.*

Cold Response 2022 will train reinforcement of northern Norway, with main action by navy and air force capacities in the Ofoten area. This region is strategically important, located about 600 kilometers from the Kola Peninsula where the Northern Fleet's nuclear submarines are based.

As the Marine Corps conducts force design, they must figure out how to supply that force. When operating in the Mediterranean, the Ospreys can fly to several support facilities. This will not be the case when operating in an environment like the North Atlantic.

This means working the logistical support challenge with the Navy to provide for afloat support and to work on pre-positioning of supplies across the arc from North America to the Baltics. How will the supply chain to support North Atlantic operations be shaped going forward?

Clearly, the renewed focus on naval integration is part of the answer. This will be a function of how the Navy reworks its own logistical support, how ashore support is built out in the region, and how the amphibious fleet is reshaped.

Maritime autonomous systems can be part of evolving support solution sets. As one participant put it: *Perhaps the supply shortfall can be mitigated by logistical movers. Having unmanned aircraft or unmanned surface vessels will undoubtedly be able to contribute going forward.*

There is clearly a shortage of amphibious shipping both in terms of combat ships and connectors for the North Atlantic mission against a peer competitor.

Another aspect being worked is how to integrate the ARG-MEU in wider fleet operations. The Marines and the Navy are working exercises in the North Atlantic to find ways to do so, and the recent BALTOPS-50 provided some insight. The evolving relationship between 2nd Expeditionary Strike Group and II MEB will clearly focus on this challenge.

An aspect of the way ahead for the ARG-MEU is its participation in fleet defense and shaping ways the amphibious force can better defend itself afloat. The F-35 has already demonstrated in the

Pacific that it can contribute significantly in this role and with the F-35 coming to 2nd MAW's operational force, it can play a similar role in the Atlantic.

Given the nature of the arc from North Carolina to the Baltics, allied F-35s will play a key role, as has already been demonstrated in BALTOPS-50 with Norwegian F-35s.

The participants indicate that they are engaged in discussions with the Navy about how to better integrate capabilities for the extended littoral operational fight. As one participant highlighted: *As the Navy focuses on integration of their fleet operations, they want to be able to use all of the assets available to them. And that is why the MEU is now part of the discussion.*

A final aspect of the potential evolution of assault support is the potential contribution of roll-on roll-off systems onboard the Osprey. This was demonstrated at last year's Deep Water exercise where MV-22 onboard capabilities allowed it to play a key role in providing C2 to a distributed force. The Marines further contended that several pertinent future capabilities are being shaped for the Osprey.

All in all, this is a good news story. Something that back in 2007, I did not even think was possible. The MV-22 Osprey is not only leading the way in combat assault support, but is a centerpiece as the Marine Corps and 2d MAW trains for operations in any clime or place.

FOURTEEN

2024 Visit: F-35s and Distributed Operations

Marine aviation on the East Coast is undergoing a significant transformation with the arrival of the F-35 Lightning II at Marine Corps Air Station Cherry Point and the continued development of the CH-53K King Stallion heavy-lift helicopter.

These platforms, combined with new approaches to distributed operations, are reshaping how the 2nd Marine Aircraft Wing contributes to American and allied defense capabilities.

The F-35 Comes to 2nd Marine Aircraft Wing: A Significant Capability for North American and European Defense

September 11, 2024

The F-35 has finally come to Cherry Point. VMFA-542 is the first of eight operational 2nd MAW F-35 squadrons, with six to be F-35B squadrons and two to be F-35C squadrons.

As 2nd Lt John Graham wrote in February 2024: *Marine Fighter Attack Squadron (VMFA) 542 became the first East Coast F-35B Lightning II Joint Strike Fighter squadron in the Fleet Marine Force to achieve initial operational capability on February 5.*

Initial operational capability means that VMFA-542 has enough operational F-35B Lightning II aircraft, trained pilots, maintainers, and support equipment to self-sustain its mission essential tasks. These include conducting close-air support, offensive anti-air warfare, strike coordination and reconnaissance, and electronic attacks.

VMFA-542 is the first operational fifth-generation squadron in II Marine Expeditionary Force, giving the aviation combat element the most lethal, survivable, and interoperable strike fighter in the U.S. inventory, said LtCol Brian Hansell, commanding officer of VMFA-542. *The F-35B is unmatched in its capability to support Marines against the advanced threats that we can expect in the future.*

The F-35 is a fifth-generation fighter jet with advanced stealth, agility and maneuverability, sensor and information fusion, and provides the pilot with real-time access to battlespace information. The F-35B Lightning II is the short-takeoff and vertical-landing variant, allowing the aircraft to operate from amphibious assault ships and expeditionary airstrips less than 2,000 feet long.[1]

The F-35s at Cherry Point will support USMC/U.S. Navy expeditionary operations in North Atlantic or Mediterranean operations, can be embedded in Nordic nations and support fleet operations, and can integrate seamlessly with the expansive fleet of F-35s flying in Europe and Israel. These aircraft can fight as a wolfpack with unprecedented data integration.

During my interviews with LtCol Hansell at MAWTS-1 in Yuma, he emphasized that the F-35 is not just another combat asset, but at the heart of empowering an expeditionary kill web-enabled force.

As Hansell explained: *During every course, we have one of the lead software design engineers for the F-35 come out as a guest lecturer. A student once asked the engineer to compare the design methodology of the F-35 Lightning II to that of the F-22 Raptor. The engineer explained that 'the F-22 was designed to be the most lethal single-ship air dominance fighter ever designed. The F-35, however, was able to leverage that experience to create a multi-role fighter designed from its very inception to hunt as a pack.*[2]'

Simply put, the F-35 does not tactically operate as a single aircraft. It hunts as a network-enabled, cooperative four-ship fighting a fused picture, and was designed to do so from the very beginning.

During my July visit to Cherry Point, Maj Carlo Bonci, executive officer for VMFA-542, highlighted the reach aspect of the F-35s as a fleet. When operating in the Nordic region during Nordic Response 24, his Marine Corps formation seamlessly integrated with British F-35s operating from HMS Queen Elizabeth, Norwegian F-35s, and other U.S. F-35s.

The author with Maj. Carlo Bonci after the visit.
Credit: 2nd MAW.

The immediate integratability of F-35s facilitated by the MADL data link was very different from the challenges of working interoperability via Link 16 with coalition partners. With F-35s, integratability was built in with MADL data links, expanding the reach of their own F-35s while contributing to other F-35s in operation.[3]

This is a major advantage in the European theater, where Norwegian, Finnish, Danish, British, Canadian, Belgian, Dutch, German, Polish, Italian, Swiss, and Romanian F-35s will provide a common operating picture over the European theater.

The F-35C model has advantages over the B model in terms of carrying more fuel and weapons, but the short takeoff and vertical landing capabilities of the B provide clear advantages for operating where traditional airfields are unavailable. In Norwegian conditions, while their Air Force uses large drag chutes to slow down for landing on icy runways, the F-35B can slow itself down and land vertically, allowing operation in icy conditions.

The Nordic engagement allows Marines to operate in two ways: tapping into protected shelter structures that Nordic Air Forces use, or operating from various locations with distributed fuel support (KC-130J or CH-53K) and land, fuel, and fly away.

Maj Bonci, a longtime Harrier operator including MEU operations, is now part of the first operational F-35 squadron on the East Coast. As the Harrier phases out of the Marine Corps, he's positioned to make Marine Corps aviation history in the standup and operation of F-35 squadrons, contributing to the reworking of American and allied defense capabilities.

When viewing the single squadron now at Cherry Point, one must look beyond into the not-too-distant future. Even though Marine Corps F-35s have had almost a decade of operational experience in the Pacific, the East Coast F-35s are arriving when a new chapter in U.S. and European defense can be written.

EABO and Reworking Aviation Ground Support: The View from 2nd Marine Wing

August 2, 2024

As the Marines rework how they approach distributed operations, a key focus is upon how to shape what they call Expeditionary Advanced Operations or EABO. This requires reworking how the air and ground elements operate together to shape a more effective distributed force with reduced force signature and an ability to operate at the point of interest more rapidly and effectively.

The Aviation Ground Support (AGS) element of the air wing is of enhanced importance in such operations, but also faces significant challenges in being able to shape the infrastructure for such operations as well.

When I have visited MAWTS-1 over the past few years as the EABO concept has been worked, the AGS personnel I have talked with believed that their role is enhanced and calls for its inclusion as the 7th function of Marine Corps Aviation.

A key element for an evolving combat architecture clearly is an ability to shape rapidly insertable infrastructure to support Marine air as it provides cover and support to the Marine Corps ground combat element. This can be seen in the reworking of the approach of the Aviation Ground Support (AGS) within MAWTS-1 to training for the execution of the Forward Air Refueling Point mission.

During a visit to MAWTS-1 in early September 2020, I discussed this evolution with Maj Steve Bancroft, Aviation Ground Support (AGS) Department Head, MAWTS-1, MCAS Yuma. In this discussion it was clear that the rethinking of how to do FARPs was part of a much broader shift in combat architecture designed to enable the USMC to contribute more effectively to blue water expeditionary operations.

The focus is not just on establishing FARPs, but to do them more rapidly, and to move them around the chess board of a blue water expeditionary space more rapidly. FARPs become not simply mobile assets, but chess pieces on a dynamic air-sea-ground expeditionary battlespace in the maritime environment.

Given this shift, Maj Bancroft made the case that the AGS capability should become the seventh key function of USMC Aviation. Currently, the six key functions of USMC Aviation are: offensive air support, anti-air warfare, assault support, air reconnaissance, electronic warfare, and control of aircraft and missiles. Bancroft argued that the Marine Corps capability to provide for expeditionary basing was a core competence which the Marines brought to the joint force and that its value was going up as the other services recognized the importance of basing flexibility.

AGS consists of 78 MOSs or Military Operational Specialties which means that when these Marines come to MAWTS-1 for a

WTI, they come together to work how to deliver the FARP capability. As Maj. Bancroft highlighted: *The Marine Wing Support Squadron is the broadest unit in the Marine Corps. When the students come to WTI, they will know a portion of aviation ground support, so the vast majority are coming and learning brand new skill sets.*[4]

Recently, I visited 2nd Marine Wing and visited MWSS-272 to discuss how they were doing in shaping such a significant transition. These combat engineers face a significant challenge in terms of working a different kind of mobile operations and will need new equipment and investments in order to do so.

During my July 2024 visit to New River, I talked with Maj Thomas Cofer, the operations officer at the squadron. Cofer has served in Iraq with the Marines and the Army in Afghanistan and is an experienced combat engineer working in the squadron to help shape its way ahead.

Throughout the discussion with Maj Cofer, he underscored the clear need to reshape their support capabilities to enable distributed operations. This is how he put it: *You have to have support that can be tailor made to deploy and project power forward.*

This means revamping the force to have support equipment which can be moved more rapidly than legacy gear. The Marine Corps has introduced new equipment and technology to the MWSS that facilitates their ability to accomplish their mission quicker and more efficiently. However, replacing earth moving equipment is difficult due to the nature of its mission.

As he put it: *Some of our support equipment are too heavy to be transported. We have recently done an exercise in the Bahamas for Distributed Aviation Operations Exercise, and we had to scale some of our planning back because of the limited ship-to-shore connectors available, and the weight of some of the support equipment. We need to be able to provide support with a lighter and more mobile package. I think that's going to be the key to success moving forward.*

Visiting MWCS-28: C2 for Distributed Aviation Operations

August 18, 2024

On July 9, 2024, I had the opportunity to visit MWCS-28 which in English is Marine Wing Communications Squadron 28. The squadron provides expeditionary communications for the aviation combat element of the II Marine Expeditionary Force. They are based at Marine Corps Air Station Cherry Point and fall under the command of Marine Air Control Group 28 and the 2nd Marine Aircraft Wing.

I was privileged to have a robust representation of the squadron and an opportunity to have a wide-ranging discussion of their approach and their recent engagement in the Nordic region in support of the Wing's most recent Arctic exercise.

Present at the discussion were Maj Mikel Santiago, operations officer, LtCol Craig Schnappinger, commanding officer, MWCS-28, MGySgt Marcus Jackson, communications chief, MWCS-28, Maj. Richard Wise, Deputy Assistant Chief of Staff, 2nd MAW G-6, 1st Lt Kara Paradowski, communications officer, MWCS-28, and 1st Lt Bennett Lax, communications officer, MWCS-28.

Speakers emphasized the importance of understanding new technologies and integrating them into existing command structures. They also discussed the challenges of modernizing military infrastructure and investing in mobile capabilities. Additionally, they highlighted the significance of interoperability, communication, and redesigning Command and Control systems to accommodate small form factors and user-friendly interfaces.

And a key point was that the Marines used various COTS equipment, but as they desire to use such equipment goes up notably in the C2 area, it is imperative to ensure proper training and maintenance procedures are in place for the deployed personnel to have the ability to make full use of the equipment in the most effective manner possible.

The clear purpose of working distributed forces is to enhance survivability but this has to be done within the logic of what is feasible. So this means discrete numbers of movement with the ability of C2 to support such movement.

The author with MWCS-28 after the discussion.
Credit: 2nd MAW.

This was exercised during Nordic Response 24, and as one Marine put it: *We utilized existing Marine Corps assets and commercial off the shelf assets in order to make ourselves more survivable and less prone to being identified hiding within the noise, so to speak. We were operating within existing infrastructure and bases that are already owned and operated by the Norwegians and working closely with them as well.* In other words, the Marines were working distributed aviation operations but within a Nordic/NATO operating space.

But the integration piece within the Nordic operating space is a work in progress and part of the evolution of C2 capability in general which is required going forward. DAO in the Nordic case to be most effective is not a pickup piece but a planned and thought through concept of operations piece, and in my view, something 2nd MAW can excel at as the Nordics work their defense integration efforts.

There is no part of the world with more experience at operating a protected and distributed force than the Nordics and coming from North Carolina, the 2nd MAW Marines can contribute and learn how to seamless integrate into regional distribution of forces. The challenge is significant as well as the opportunities for innovation.

The CH-53K Progressing on Pace: The Perspective from HMH-461, MCAS New River

August 5, 2024

During my recent visit to 2nd Marine Aircraft Wing at Marine Corps Air Station New River, I met with two officers who have been involved from the outset in HMH-461 standing up the CH-53K King Stallion. Their insights reinforced a key point I've long argued: this aircraft represents a fundamental transformation, not just an incremental improvement.

To the casual observer, the Super Stallion (CH-53E) and the King Stallion (CH-53K) look like the same aircraft. This similarity creates a challenge in understanding how radically different these platforms truly are. As I wrote in 2020, if it were called the CH-55, perhaps people would better grasp that these are very different air platforms with very different capabilities. While they share a similar logistical footprint by deliberate design, allowing them to operate off amphibious ships, the fundamental difference is this: the CH-53E is a mechanical aircraft, while the CH-53K is digital from the ground up.

The CH-53K is faster, carries more kit, can distribute its load to multiple locations without landing, and leverages its digitality for significant advancements in maintenance, task force operations, updates, and potential integration with unmanned systems. These capabilities create a fundamentally different lift platform with strategic implications for force mobility across the combat spectrum.

During my July 2024 visit, I spoke with Capt Jeffrey Stanton, assistant operations officer, and Capt Philip Wood, CH-53K pilot and pilot training officer. Both are legacy heavy lift operators who joined the squadron simultaneously and have been on the ground floor of CH-53K operations.

The CH-53K is a completely different aircraft from the CH-53E, Capt. Wood explained. *The way you physically fly it, the way you plan for operations, and the way you maintain it are completely different.* He emphasized that new pilots are specifically told not to treat it as an Echo but to change their mindset entirely.

A U.S. Marine Corps CH-53K King Stallion helicopter, assigned to Marine Heavy Helicopter Squadron (HMH) 461, conducts an external lift at Auxiliary Airfield II near Yuma, Arizona, March 28, 2023. (U.S. Marine Corps still image extracted from video by Cpl. Jaye Townsend).

It is more of mental than physical game in operating the aircraft, Wood continued. *You are focused on manipulating everything the aircraft can do. You are focused outside of the aircraft on what the pilots can do to support operations. The pilots have much more situational awareness and can operate the aircraft to support the changing operations environments more rapidly.*

The aircraft's precision hover capability was dramatically demonstrated during the recovery of a downed Navy MH-60S Seahawk from the bottom of a ravine in California's Inyo National Forest. This operation required the exact kind of precision that sets the CH-53K apart from its predecessor.

As LtCol Frank noted during my 2020 visit to VMX-1, the aircraft's hover capability is revolutionary: *We have the ability to hit position hold in the 53K and have the aircraft maintain pretty much within one foot of its intended hover point, one foot forward, lateral and AFT, and then one foot of vertical elevation change. It will maintain that hover until the end of the time if required.*

The author with Capt. Jeffrey Stanton and Capt. Philip Wood after the discussion. Credit: 2nd MAW.

Capt Stanton, who was part of the ground crew during the Seahawk recovery, emphasized how the King Stallion facilitated the operation: *Using the CH-53K would reduce significantly the risk factors involved in such an operation. It was around a 2-minute precision hover to come in and allow the helicopter support team to rig the Seahawk.*

The aircraft is progressing through its developmental phases, with new capabilities being released as they're certified. Currently, the King Stallion is largely limited operationally to CH-53E-type missions, but even these are being performed more efficiently. A prime example is the ability to carry a JLTV ashore in one sweep—something the CH-53E cannot do as easily or efficiently.

As Col Fleeger, the NAVAIR program manager of H-53 helicopter programs, noted in a recent interview I did with her: *The operating crews will drive the out of the box thinking about how we can use our heavy lift assets to do new things and work new thinking about what payloads we can and should carry.*[5]

Both officers I spoke with fully agreed with this perspective. As Capt Wood put it: *There are a lot of things we could do now with the CH-53K that have yet to explore. And there are certainly things the aircraft can do that we have not even thought of. The fleet pilots will come up with new ways of doing things and employing its new capabilities.*

The CH-53K represents more than technological advancement.

It's a paradigm shift in vertical heavy lift capabilities. Its digital architecture, precision hover, enhanced payload capacity, and operational flexibility position it as a force multiplier in an era where strategic mobility defines tactical advantage. As capabilities continue to be certified and released, we can expect to see innovative applications that fully leverage this remarkable platform's potential.

FIFTEEN

2025 Visit: Working Distributed Air Operations

As global military challenges evolve, the U.S. Marine Corps is fundamentally reimagining its aviation operations to maintain battlefield superiority in future conflicts. Distributed Aviation Operations and operations from forward basing options, afloat or ashore are increasingly the focus of attention.

My 2025 visit focused on these developments plus getting updates on core USMC air systems.

Distributed Aviation Operations: Marine Wing Support Squadron Adapts for Future Combat

In an increasingly complex and contested battlefield environment, the U.S. Marine Corps is adapting its aviation support strategy to meet emerging challenges.

Captain Medlen, an eight-year Marine Corps veteran currently commanding the Engineers Company of Marine Wing Support Squadron (MWSS) 272, provided valuable insights into how distributed aviation operations are evolving based on recent deployment experiences.

I met with Captain Medlen at his office in New River on 30

April 2024 to discuss the approach and the recent exercise in the Bahamas.

The author with Captain Medlen after the discussion. Credit: 2nd MAW.

We need to shift towards single touch points with different things, explains Captain Medlen, contrasting the old model with what's needed for future conflicts. *We think of old school FARPs [Forward Arming and Refueling Points] in Afghanistan and Iraq as large FOB [Forward Operating Base] style, Walmart supercenters with everything under one roof. We need to start looking at Mom and Pop type stores - this place has gas, this place has ordnance, this place has somewhere to sleep.*

This distributed approach stems from a fundamental warfighting concern: survivability. As potential adversaries develop increasingly sophisticated targeting capabilities, centralizing resources creates vulnerable nodes that can be easily compromised.

By distributing capabilities across multiple smaller sites, the Marines create a more resilient network of mission critical assets and personnel. The Captain underscored: *If I lose one of maybe 20 parking spots, that's not the end of the world,* Medlen notes. *You can hurt a small part of us, but not necessarily destroy the body.*

MWSS 272 recently deployed to the Bahamas to test these concepts in a real-world environment. Using New River/Camp

Lejeune as their primary staging area (Echelon 4), they established a command-and-control outpost at the U.S. Navy's Atlantic Undersea Test and Evaluation Center (AUTEC) on Andros Island (Echelon 5). From there, they pushed teams to uninhabited islands with little to no available infrastructure in place.

The deployment demonstrated the squadron's ability to create rapidly operating locations from scratch. At "Site Four," combat engineers and heavy equipment operators removed 30-40 foot trees and transformed the area into a functional two-point landing zone in just under one week. At "Site Three," they executed a proof-of-concept operation using only hand tools and chainsaws to create a single-point landing zone.

Critically, these sites served different purposes. While one might provide fuel, another might offer overnight maintenance capabilities. This approach creates operational flexibility while complicating enemy targeting efforts.

A key component of distributed aviation operations is self-sustainability. Captain Medlen highlights how the Marine Aircraft Wing (MAW) can support itself in littoral environments. *We can make potable water from seawater at the main AUTEC installation, package it, put it on a CH-53, add some boxes of chow and a little bit of ordnance. When they go get gas somewhere, they drop all that stuff off. Now we have self-supporting sustainment internal to Second MAW.*

This capability for aerial resupply enables Marines to maintain a distributed posture without relying on vulnerable ground logistics chains or large fixed bases.

Despite promising progress, several challenges remain. Captain Medlen identified command and control integration between ground and air elements as a significant hurdle, noting that many Marines, including himself, are experiencing their first assignment in aviation units. *Being able to bridge the gap in what we understand as C2 from the ground side, and the C2 capabilities that the aircraft and the pilots are familiar with, and being able to make it more cohesive and reduce uncertainty is something that we're working on getting more proficient at,* he explains.

The recent introduction of Marine Air-Ground Tablets (MAGTABs) (digital tablets for sharing operational information) has improved this situation, allowing pilots to quickly access information about distributed sites, including available resources, munitions types, and fuel quantities.

Another critical challenge is deploying ground support elements. Unlike aviation units that can self-deploy, ground support elements require transportation. Medlen emphasizes the need to "integrate early and often with the flying squadrons" to ensure ground capabilities arrive where needed.

Perhaps most significantly, Medlen identifies distributed medical support as imperative for future conflicts. During the Bahamas deployment, rehearsals and clearly established medical capabilities proved to be instrumental in ensuring that combat readiness was maintained. *Do we have general consolidation points for routine casualties? Somewhere else for surgical capabilities? Another area for imagery?* Medlen asks, pointing out that distributing medical assets prevents the enemy from targeting a single medical facility, which could have devastating psychological impacts on forces.

The experience in the Bahamas demonstrated that a Marine Wing Support Squadron could rapidly establish multiple landing zones in austere environments, creating a distributed network of capabilities that enhances operational reach while minimizing vulnerability.

Captain Medlen envisions continuing this approach in future operations: *If we're there for a longer time, I'm not creating two landing zones. I'm creating as many as possible throughout the operational area.*

This distributed aviation operations concept represents a significant shift from recent conflicts characterized by large, established bases.

By adapting to more distributed postures, the Marine Corps aims to maintain operational effectiveness while reducing vulnerability in contested environments where precision munitions could quickly compromise centralized facilities.

As Captain Medlen notes, this approach creates "standoff" and "uncertainty for the enemy" which are essential elements of advantage in future conflicts.

The KC-130J: A Key Enabler of Marine Corps Aviation Operations

U.S. Marines with Marine Medium Tiltrotor Squadron (VMM) 261 receive fuel in an MV-22B Osprey from a KC-130J Super Hercules with Marine Aerial Refueler Transport Squadron (VMGR) 252 over the Atlantic Ocean near North Carolina, May 6, 2025. (Photo by Lance Cpl Mya Seymour).

On 30 April 2025, I interviewed Major White, the Operations Officer at VMGR-252, a KC-130J squadron based at Marine Corps Air Station Cherry Point and part of 2nd Marine Aircraft Wing. Unlike other Marine Aircraft Wings, 2nd MAW serves as a global support force, requiring squadrons like VMGR-252 to operate worldwide at a moment's notice.

In modern military operations, few platforms play as vital a role in Marine Corps aviation as the KC-130J. Major White, an experienced KC-130J officer with multiple deployments and Weapons and Tactics Instructor qualifications, revealed how this aircraft serves as the linchpin for Marine expeditionary capabilities worldwide.

Our reach is unlimited, Major White explains. *We can get there faster than pretty much anything else.* This self-deployable capability sets the KC-130J apart from other platforms that often require additional support to reach distant operational areas. The KC-130J's dual-role capability makes it particularly valuable, serving as both a critical air refueling platform and tactical transport asset, though this creates inherent tensions in resource allocation.

The air refueling function is especially vital for CH-53 helicopters. *The CH-53s use us a lot because to my knowledge there's no contract civilian tanker that can fly slow enough,* Major White explains. *We really work hard to make sure we're able to sustain them.*+

This tanking capability becomes increasingly important as the Marine Corps moves toward distributed operations. The ability to extend rotary-wing asset range through aerial refueling directly enables the distributed force posture envisioned in current Marine Corps doctrine.

The KC-130J squadron at Cherry Point benefits from proximity to the new CH-53K squadron at New River, enabling experience with the new generation heavy lift helicopter. The CH-53K's intermodal cargo system allows direct transfer of Air Mobility Command 463L pallets from fixed-wing transport aircraft without reconfiguration, with an internal pallet locking system eliminating cargo straps and significantly enhancing cargo operation speed.

As described by 2nd MAW: *The tail-to-tail transfer of supplies allowed distribution of sustainment in the minimum time period of vulnerability by reducing break-bulk requirements. U.S. Marines with 2nd Marine Aircraft Wing experimented with dynamic, assault-support capabilities in a distributed-aviation environment.*[1]

Distributed operations highlight the growing challenge of sustainment in force deployment. If forces are distributed, how long, where, and how do you keep them supplied and their equipment operational?

Major White acknowledged the challenges of balancing the KC-130J's dual roles: *There's tension between the tanking function, which is important, and the lift function."* This becomes particularly critical when considering distributed operations, which "create an exponential increase in the problem of sustainability."

U.S. Marines with Marine Aerial Refueler Transport Squadron (VMGR) 252 transfer supplies from a KC-130J Hercules into a CH-53K King Stallion assigned to Marine Heavy Helicopter Squadron (HMH) 461 with an extended boom forklift – military millennium vehicle assigned to Marine Aviation Logistics Squadron (MALS) 29 at Marine Corps Air Station New River, North Carolina, June 7, 2023.

The squadron works with HIMARS rapid deployment, quickly transporting this critical weapon system over significant distances despite payload limitations.

Like many Marine Corps platforms, the KC-130J faces modernization challenges. Major White highlighted that some aircraft are 15-20 years old with outdated software compared to Air Force equivalents. *We've bought a system, but the Air Force already upgraded to a better system, so we're still behind where we could be,* he explained.

This points to a larger defense acquisition issue: the focus on new platforms often overshadows the critical need to maintain and upgrade existing systems. As Major White notes, it's not a flying problem but a supply problem.

The author with Major White after the discussion.
Credit: 2nd MAW.

Despite challenges, Major White remains confident in the squadron's mission capability. *We're Marines, and we'll find a way to get the mission done,* he stated, emphasizing how integrating smart, creative personnel helps overcome resource limitations.

This integration is enhanced through Marine Aviation Weapons and Tactics Squadron training cycles, which Major White described as crucial for cross-platform learning and innovation. *I think from every iteration, we learn from each other and make ourselves better.*

However, as distributed operations become increasingly central to Marine Corps strategy, the question remains whether logistics and sustainability will receive the attention and funding they require.

As Marine Corps operations continue evolving toward distributed concepts, sustainment capabilities provided by platforms like the KC-130J will remain essential. Major White's insights reveal both the impressive capabilities these aircraft bring and the ongoing challenges in keeping them ready for tomorrow's conflicts. In this context, the KC-130J stands as both a solution to current operational needs and a symbol of the broader challenges facing military logistics in an era of great power competition and distributed operations.

The Osprey Evolution: From Assault Aircraft to Multi-Mission Platform

By Robbin Laird

The MV-22 Osprey has come a long way since its early days as a novel tiltrotor aircraft. What began as a specialized assault platform has evolved into an incredibly versatile multi-mission aircraft capable of transforming Marine Corps operations.

In an interview held in his office at Marine Corps Air Station New River on April 30, 2025, Major Sean Timothy Penczak, Operations Officer with VMM-162, shared his extensive experience with the platform and provided insights into its evolving capabilities.

Major Penczak joined the Marine Corps in January 2009 and, after working briefly as a Logistics Officer, made his way to flight school. In 2012, he selected the MV-22 at a time when there were reportedly "less than 100 Osprey pilots in the fleet." His career has included deployments to Afghanistan, Kuwait for Operation Inherent Resolve, Spain with the Special Purpose MAGTF Crisis Response Africa, Hawaii, and Australia.

During his 2014 deployment to Afghanistan, Major Penczak participated in a pilot program called "Enhanced Casevac," which demonstrated one of the Osprey's unique advantages. While traditional helicopters were limited to a 40 nautical mile "golden hour" range, the MV-22 could reach that distance and return in just 20 minutes.

I could get up off deck in less than 15 minutes, out to pick somebody up, and home in about 10 to 15 minutes, Penczak recalled. *You can max blast out to get to a point of injury.*

This capability proved invaluable for casualty evacuation, allowing injured personnel to be transported to medical facilities within the critical "golden hour" timeframe. The aircraft's unique combination of helicopter landing capability with fixed-wing speed and range made it ideal for this mission.

The author with Major Penczak after the discussion. Credit: 2nd MAW.

One of the Osprey's most revolutionary capabilities is its self-deployment range. Major Penczak described missions in Spain where the aircraft could *self-deploy to support units located in a lot of different areas, hopping between islands across the Mediterranean without requiring the extensive logistical support needed by other aircraft.*

In Australia, where Penczak participated in two Marine Rotational Force-Darwin deployments, the Osprey demonstrated similar advantages. *We could self-deploy ourselves all over Australia,* he noted, explaining how a journey that would take multiple hops for our H-1 brethren could be completed in a single hour-long flight by the MV-22.

As the interview progressed, the discussion turned to the "payload revolution." The Osprey has evolved from primarily transporting Marines to supporting a much wider range of mission sets through various payloads and configurations.

What do you want to put on the bird to utilize? Penczak asked rhetorically.

The aircraft can be configured for numerous roles:

- Casualty evacuation with medical teams and equipment.
- Command and control platform coordinating other aircraft.
- ISR (Intelligence, Surveillance, Reconnaissance) platform.
- Logistics support moving critical supplies.
- Troop transport for rapid insert/extract operations.

Major Penczak highlighted a recent exercise in the Bahamas where the Osprey demonstrated its exceptional transport capabilities. The platform efficiently moved an individual who required medical treatment from a remote island to Miami in under 40 minutes – a rapid transit performance that distinguishes the Osprey from other aviation platforms.

Looking to the future, a significant enhancement which I believe is important for the Osprey is the implementation of a new digital backbone. This upgrade would allow greater flexibility in integrating various payloads without extensive integration with the airplane's systems, enabling operators to be more responsive to mission needs.

The less computers, less weight, Penczak posited, noting how the incorporation of smaller, more advanced systems could further enhance the aircraft's capabilities while reducing physical constraints, enabling a variety of different payloads for different missions.

Operating such a versatile platform presents unique training challenges. Major Penczak explained the deliberate and progressive approach to pilot development, with a minimum of approximately 500 hours of flight time required before pilots can become aircraft commanders.

The training pipeline builds from basic flight skills to advanced tactical employment, including night operations, low-altitude tactics, aerial refueling, and specialized mission sets like hoist operations. Simulation plays a crucial role in this process, allowing pilots to practice complex scenarios before executing them in the aircraft.

The interview concluded with a discussion of how the Osprey

supports the concept of Distributed Aviation Operations (DAO), particularly in potential island-chain scenarios. The MV-22's ability to operate from austere locations without requiring established runways makes it invaluable for this operational concept.

Marines aren't going to sit at an airfield. That's just not what Marines are supposed to do, Penczak noted. The Osprey can connect these distributed forces, delivering personnel, equipment, and supplies to locations inaccessible to fixed-wing aircraft or surface connectors.

From its early days as a specialized assault platform to its current role as a multi-mission enabler, the MV-22 Osprey has proven itself as a transformative capability for the Marine Corps.

As Major Penczak put it, *the Osprey represents the best of both worlds, combining the vertical landing capability of a helicopter with the speed and range of a fixed-wing aircraft.*

As the platform continues to evolve with new technology and tactics, its central role in Marine Corps operations seems assured in the period ahead.

This interview is good place to end the book, as it highlights the path I have seen at 2nd Marine Air Wing since my first encounter with them in 2007.

About the Author

Dr. Robbin F. Laird is a long-time analyst of global defence issues. He has worked in the U.S. government and several think tanks, including the Center for Naval Analyses and the Institute for Defense Analyses.

The author onboard an Osprey with other guests to view the USS Wasp sea trials of the F-35B in 2025.

He is a frequent op-ed contributor to the defence press, and he has written several books on international security issues. He has followed the path of the introduction and modernization of a number of USMC air systems, including the Osprey, the F-35 and the CH-53K.

He has been involved for a number of years in issues involving the emergence of the maritime kill web and the significant impact of the emergence of maritime autonomous systems, a development critical to evolving USMC operational capabilities.

He is the editor of two websites, *Second Line of defence* and *defence.info*.

He is a member of the Board of Contributors of *Breaking defence* and publishes there on a regular basis.

He is a research fellow with The Sir Richard Williams Foundation.

He regularly travels throughout Europe and conducts interviews with leading policymakers in the region.

A further note from the editor about his career and his approach to analysis:

Over the course of my career, I have focused on military and security affairs, with an emphasis on adapting to the ongoing changes in global defense policy. My professional background includes teaching positions at Columbia University, Queens College, Princeton University, and Johns Hopkins University, along with work with Dr. Brzezinski at the Research Institute of International Change. These early roles gave me the opportunity to learn from and work with colleagues committed to international security issues, both academically and operationally.

My interest in European security and the Soviet Union began during the Cold War and helped shape my later work. Then with the collapse of the Soviet Union, in the early 1990s, I visited Ukraine, Russia, and Belarus and participated in one of the initial Pentagon studies examining the newly independent Slavic states. These experiences reinforced for me the importance of trying to anticipate geopolitical developments and contributed to the way I approach strategic analysis.

Over time, I have sought to bridge the gap between policy discussions and operational realities. I have worked with organizations such as the Center for Naval Analysis and the Institute for Defense Analysis, experiences that contributed to my understanding of the evolving nature of contemporary defense challenges.

Recognizing the changing landscape of how defense information is shared, I co-founded Second Line of Defense in 2010 and Defense.info in 2018. My intention has been to provide platforms for honest and informed discussion among professionals interested in defense policy, technology, and operations. These aren't just websites but they're strategic communities where policy, technology, and operational realities converge

I continue to write regularly as well for outlets like Breaking Defense, aiming to offer context and perspective on ongoing transformations in defense. My approach has always been to draw on historical context, technological developments, and the insights of those working directly in the field. I try to avoid taking conventional wisdom at face value and instead focus on the practical challenges and opportunities facing modern military organizations.

My work has given me the chance to learn from colleagues and policymakers in Europe, North America, and Australia, and to develop a broader view of the issues at stake in defense modernization and strategic policy choices in a world in flux.

Operating under the motto "If everyone is thinking alike, someone isn't thinking" — borrowed from General Patton -- I have built my career on challenging conventional wisdom and pushing defense communities toward innovative solutions. My approach combines historical perspective with technological fluency and operational understanding, a combination that is particularly valuable in an era of rapid strategic change.

Overall, my goal is to contribute constructively to assessments about the future of defense policy at a time of significant change, emphasizing the need for both innovation and a solid understanding of past experience and its continuing impact.

About the Contributors

Edward Timperlake

Edward Timperlake's professional journey stands as a powerful example of multidimensional commitment to national defense, government leadership, and investigative analysis. With a career spanning high-stakes operational roles, senior government appointments, and insightful contributions to defense literature, Timperlake's story demonstrates a rare blend of practical experience and strategic thought leadership.

Timperlake began his career with a strong academic foundation, graduating from the United States Naval Academy in 1969. He pursued further education by earning an MBA from Cornell University, preparing himself for both operational leadership and complex organizational roles.

Entering the U.S. Marine Corps, Timperlake trained as a fighter pilot—with particular expertise in carrier operations—and quickly distinguished himself. His service included a combat tour during the Vietnam War, where he was awarded the Vietnam Service Medal (2 stars). Rising through the ranks, he commanded VMFA-321, a Marine Corps reserve fighter squadron, highlighting his leadership of operational aviation units and engagement in advanced tactical training programs.

After his active military service, Timperlake transitioned to several senior government positions:

As Assistant Secretary of the U.S. Department of Veterans Affairs (1989–1992), he played pivotal roles in Congressional, public, and intergovernmental affairs, including leading medical mobilization during Operation Desert Shield/Desert Storm and acting as a key spokesperson on Gulf War Illnesses.

Serving as Principal Director of Mobilization Planning and Requirements in the Office of the Secretary of Defense under President Reagan, he contributed to both DOD mobilization planning and continuity of government initiatives.

On Capitol Hill, Timperlake worked with the House Committee on Rules, focusing on national security issues including the investigation of illegal foreign contributions to political campaigns and participating in NATO's North Atlantic Assembly affairs.

Transitioning into the private sector, he directed classified studies on military modernization as a Program Manager at The Analytic Sciences Corporation (TASC). Notably, he developed the TASCFORM methodology, a significant innovation for assessing global military air power modernization—of lasting value to sponsors such as the Department of Defense and the CIA.

Timperlake's expertise reached a wider audience through his editorial leadership and authorship:

As key part of the leadership team for *Second Line of Defense* and *Defense.info*, he contributed in-depth reporting and analysis on defense technology and policy issues.

His books have garnered national attention, including the New York Times best-selling *Year of the Rat*, *Red Dragon Rising*, *Showdown*, and *Rebuilding American Military Power in the Pacific: A 21st Century Strategy*, co-authored with Robbin Laird and Richard Weitz.

Demonstrating a commitment to public service beyond government, he has served on the board of the Vietnam Children's Fund, facilitating the construction of elementary schools in Vietnam, a testament to his long-term vision for reconciliation and community-building.

Edward Timperlake's varied career highlights a unique fusion of combat experience, strategic planning, policy innovation, and a passion for informing public understanding. His influence continues to resonate in U.S. security circles, military modernization debates, and humanitarian efforts making him a respected voice in every sphere he has engaged.

Murielle Delaporte

Murielle Delaporte is a defense analyst based in France and the United States. She has worked in the French Government and various think tanks in both France (*Institut français des relations internationales* in Paris) and the United States (East West Institute for Security Studies in New York).

She received three political science and security studies degrees from Sciences Po and Sorbonne in Paris, as well as from Georgetown University in Washington, D.C.

She has published several books and articles in both French and English on defense and strategic issues. After being a correspondent for various publications (TTU in France, Military Logistics International in the UK) and an occasional contributor to others (*FrontLine Defence* in Canada, and *Leatherneck* in the United States), she is a regular contributor to *Breaking Defense* since 2013 and a guest lecturer at Sciences Po – Bordeaux since 2017.

She is currently the editor-in-chief of a French magazine and website specialized on military and security issues – https://operationnels.com- which she founded in 2009.

In the course of her work, she has visited several U.S. and Canadian military bases and has reported from Mali, Afghanistan, Djibouti and other operational areas with the French Armed Forces.

About Our Websites

Sixteen years ago, we launched *Second Line of Defense* based on General Patton's warning that "if everyone is thinking alike then no one is thinking" wasn't philosophical wisdom for it is operational necessity. When defense establishments fall into intellectual lockstep, people die. When military leaders embrace comfortable assumptions rather than uncomfortable realities, they prepare for yesterday's wars while tomorrow's conflicts explode around them.

This isn't academic theorizing. This is life-and-death analysis for decision-makers who cannot afford to get it wrong.

From our inception, we established three non-negotiable analytical pillars: understanding how the global security system actually evolves (not how we wish it would), examining how military force operates within that brutal reality (not within sanitized war games), and tracking emerging capabilities that fundamentally alter warfighter effectiveness (not distant science fiction). These are not research projects; they were survival imperatives for liberal democracies facing adversaries who don't play by Western rules.

The defense analytical community has developed a pathological addiction to consensus-building that actively undermines strategic thinking. Academic institutions reward scholars who build on established frameworks rather than challenge them. Government agencies promote analysts who validate existing programs rather than question their effectiveness.

The result is an echo chamber that consistently misses the most important developments while obsessing over the least relevant ones.

We made a deliberate decision to reject this intellectual corruption. We focus on what can be done rather than what sounds sophisticated, on real options for decision-makers rather than theoretical constructs that impress peer reviewers, and on the paths actually taken by military organizations under pressure rather than the paths recommended by consultants who have never faced enemy fire.

This approach required building relationships with a different kind of analytical network, practitioners, innovators, and strategists who understand that effective defense analysis demands both ruthless criticism and constructive problem-solving. We sought out multi-dimensional thinkers because multifaceted problems cannot be solved by narrow specialists who have never ventured outside their academic silos.

The defense establishment too often suffers from chronic platform fetishism, an obsession with individual weapon systems divorced from their operational context. Our analytical approach extends beyond this narrow focus to examine how new technologies enable entirely new concepts of operations. For example, the tiltrotor aircraft revolution didn't just provide the military with a new platform because it opened tactical and operational possibilities that demanded fresh thinking about mission planning, maintenance logistics, pilot training, and force structure design. Understanding the aircraft meant nothing without understanding the system it enabled.

This systems-thinking approach proved invaluable for understanding the broader implications of military innovation, particularly the second- and third-order effects that defense planners consistently underestimate. New capabilities don't exist in isolation: rather they interact with existing systems, create new vulnerabilities even as they address old ones, and produce cascading effects that weren't visible during initial development phases.

The drone revolution exemplifies this dynamic complexity with devastating clarity. What began as simple remotely piloted vehicles for intelligence gathering evolved into a comprehensive ecosystem of unmanned systems spanning air, ground, and maritime domains. But more importantly, it fundamentally altered the strategic balance between state and non-state actors, democratized precision strike capabilities, and collapsed the traditional distinctions between tactical and strategic effects.

One of our most important analytical innovations has been our relentless emphasis on the "fight tonight" force rather than speculative 2040 capabilities. This perspective stems from a hard-learned understanding that military forces must be prepared to operate effectively with the systems they have, not the systems they wish they had or hope to develop over the next two decades.

This isn't anti-innovation bias -- it's pro-survival realism.

The approach recognizes that effective deterrence and crisis response depend on current capabilities, realistic near-term improvements, and adaptive strategies that can evolve with changing circumstances. We analyze what has been done successfully and how it can be replicated, because focusing too heavily on distant future capabilities creates dangerous gaps between current readiness and immediate threats.

The drone warfare revolution perfectly illustrates why this perspective matters. While defense establishments were planning elaborate future warfare scenarios for 2035, adversaries and proxies were using commercially available drones modified for military purposes to reshape battlefields from Ukraine to Gaza to Yemen. Military forces that adapted existing systems and tactics to counter these immediate threats proved vastly more effective than those waiting for perfect future solutions that might never arrive.

The lesson is clear: adaptation beats anticipation, and flexibility trumps perfection when the shooting starts.

In an era of increasing polarization and institutional groupthink, maintaining analytical independence becomes not just important but existentially necessary. Effective defense analysis requires the ability to question fundamental assumptions, challenge established practices and consider alternative perspectives without losing sight of practical constraints and operational realities.

As we continue this analytical journey, we remain committed to the principles that have guided us from the beginning: rigorous analysis over convenient consensus, practical focus over theoretical elegance, forward-thinking perspective over bureaucratic inertia, and intellectual independence over institutional approval.

The strategic environment will continue evolving in ways that surprise the consensus and vindicate the skeptics. New technologies will emerge that reshape military possibilities faster than procurement systems can adapt. Fresh challenges will arise that demand intellectual agility rather than bureaucratic momentum.

Our mission remains constant: providing insights that help decision-makers navigate complexity and uncertainty in service of effective defense and deterrence. We analyze the world as it is, not as we wish it were. We focus on capabilities that exist, not capabilities that might exist someday. We examine strategies that work, not strategies that sound impressive in briefing rooms.

The next few years promise to be at least as dynamic as the last sixteen, with the added urgency that comes from adversaries who have learned to exploit Western analytical blind spots. We look forward to continuing this analytical journey, tracking the evolution of military capabilities and strategic realities, and contributing to the broader conversation about how liberal democracies can maintain security in an increasingly complex and dangerous world.

The stakes have never been higher. The need for independent, hard-hitting analysis has never been greater. And the cost of getting it wrong has never been more catastrophic.

October 1, 2025

USMC Transformation Series: A Comprehensive Analysis

Our USMC transformation series presents a unified argument: the Marine Corps has undergone a fundamental metamorphosis from a traditional amphibious force optimized for land wars into a distributed, multi-domain force capable of engaging peer competitors in the era of major power competition.

This transformation is anchored by three revolutionary air systems— the F-35B, V-22 Osprey, and CH-53K that together create entirely new operational possibilities.

"The U.S. Marine Corps Transformation Path" (2022) establishes the strategic framework, positioning the USMC's evolution from land war operations to peer competitor engagement

The F-35 Journey

"My Fifth-Generation Journey: 2004-2018" (2024) and its **Second Edition** (2025) chronicle the F-35's development from your personal perspective as an observer and analyst

"Three Dimensional Warriors" (2010) provides the earliest conceptual foundation, introducing how the F-35B and Osprey enable "three block war" capabilities

The Tiltrotor Enterprise

"A Tiltrotor Enterprise: From Iraq to the Future" (2025) presents the Osprey not as a single aircraft but as the anchor of a comprehensive enterprise

"A Tiltrotor Perspective: Exploring the Experience" (2025) provides the human story behind the controversial aircraft's vindication in combat

"The Role of the Osprey in the Pivot to the Pacific" (2023) connects tiltrotor capabilities to broader strategic reorientation

The Heavy Lift Revolution

"The Coming of the CH-53K" (2023) argues for reconceptualizing this as essentially a new aircraft (the "CH-55") due to its digital transformation

Training and Doctrine Evolution

"MAWTS-1: An Incubator for Military Transformation" (2024) examines how the Marines' premier training center adapts doctrine to new capabilities

"MAWTS-1: 2023 Visit and Interviews" (2023) provides contemporary insights into training evolution

Methodological Approach

Several distinctive methods are used in the series;

Longitudinal Observation: We have tracked these transformations over more than a decade, providing unique insights into how initial concepts evolved into operational reality.

Practitioner-Focused Interviews: Rather than relying solely on official doctrine or academic theory, we prioritized voices of pilots, maintainers, commanders, and other operators who live these transformations daily.

Global Perspective: We examine not just U.S. operations but how allied nations integrate these systems, providing comparative insights.

Systems Thinking: We consistently view individual platforms as components of larger enterprises rather than isolated capabilities.

Key Analytical Insights

Beyond Platform Acquisition

The boss provide insights into successful military transformation requires more than new equipment — it demands new thinking, new operational concepts, and new approaches to training and doctrine.

The Digital Transformation

A recurring theme is how these platforms represent fundamentally different capabilities due to their digital architecture, not just improved versions of legacy systems.

Coalition and Alliance Implications

We regularly examine how these transformations affect allied operations, noting that many coalition partners find the USMC a more relevant model than the larger U.S. military structure.

Operational Vindication

The books consistently demonstrate how platforms initially dismissed by critics (particularly the Osprey) proved their value through combat operations and operational use.

Evolution of Focus

The series shows an evolution in focus:

Early works (2010-2022) focused on individual platform capabilities and initial integration

Middle period (2023-2024) emphasized enterprise-level thinking and training transformation

Recent works (2025-) demonstrate maturity in understanding how these systems create new operational paradigms

Contribution to Military Analysis

The series makes several unique contributions to military transformation literature:

Longitudinal Documentation: Few analysts have tracked a transformation this comprehensively over such an extended period

Operator-Centric Perspective: The emphasis on practitioner voices provides insights often missing from policy-focused analyses

Integration Analysis: Rather than examining platforms in isolation, we consistently explore how they work together to create new capabilities

Strategic-Tactical Bridge: The books connect high-level strategic shifts to tactical implementation in ways that illuminate both levels

Significance for Understanding Military Change

Collectively, the books document not just the USMC's transformation but provide a case study in how military institutions adapt to changing strategic environments.

They demonstrate that successful transformation requires alignment between strategic vision, technological capability, operational doctrine, and training evolution are all anchored by personnel who understand and embrace change.

The series stands as both historical documentation and analytical framework for understanding military transformation in the 21st century.

Notes

2. First Encounter: The USMC on HMS Illustrious, 2007

1. https://commons.wikimedia.org/wiki/File:USMC_Harriers_line_the_deck_of_HMS_Illustrious_MOD_45147594.jpg

6. 2011 Visit to New River: USMC Con-Ops in Evolution

1. https://abcnews.go.com/International/us-fighter-jet-crashes-benghazi/story?id=13191505
2. https://defense.info/defense-systems/the-libyan-trap-mission-the-osprey-makes-a-key-combat-impact/
3. https://defense.info/defense-systems/the-libyan-trap-mission-the-osprey-makes-a-key-combat-impact/
4. https://www.csmonitor.com/USA/Military/2011/0322/How-an-MV-22-Osprey-rescued-a-downed-US-pilot-in-Libya

8. 2013 Visit: Ospreys, Harriers and the Future

1. In the United States Marine Corps, a "named mission" typically refers to a specific, designated operational responsibility assigned to a unit or command. These missions often have formal designations and reflect particular capabilities, areas of responsibility, or operational focuses within the Marine Corps structure.
2. Brunei Darussalam Time (BNT) is 8 hours ahead of Coordinated Universal Time (UTC). This time zone is in use during standard time in: Asia.

9. 2014 Visit: Reshaping USMC Expeditionary Capabilities

1. https://sldinfo.com/kc-130-squadron-moves-to-air-station-iwakuni/
2. https://sldinfo.com/2014/04/the-next-building-block-in-pacific-defense-the-us-philippines-defense-agreement-2014/
3. https://sldinfo.com/2013/11/we-are-ready-now-sir-the-21st-century-version/
4. https://www.marines.mil/News/Messages/Messages-Display/Article/2377486/update-to-mos-7315-and-creation-of-7318-pmos/
5. http://www.navair.navy.mil/v22/?fuseaction=faq.main
6. https://sldinfo.com/2015/02/an-update-on-the-uss-america-a-discussion-with-captain-robert-hall-february-2015/

10. 2015 Visit: Strands of Transition

1. https://www.marinecorpstimes.com/2015/01/25/some-marine-prowler-offi cers-will-soon-operate-drones/
2. http://www.marinecorpstimes.com/story/military/careers/marine-corps/offi cer/2015/01/25/marine-corps-prowler-officers-to-operate-drones/22139041/
3. https://sldinfo.com/2014/03/a-key-army-contribution-to-pacific-defense-the-evolving-missile-defense-mission/
4. https://www.defensemedianetwork.com/stories/return-to-the-sea/

11. 2019 Visits: Transitioning to Arctic Operations

1. https://defense.info/global-dynamics/2018/12/the-return-of-direct-defense-for-northern-europe-not-your-daddys-cold-war/
2. https://sldinfo.com/2018/05/trident-juncture-2018-the-defense-of-norway-and-working-21st-century-deterrence-in-depth/
3. https://defense.info/interview-of-the-week/brigadier-general-jan-ove-rygg-on-shaping-the-way-ahead-for-the-norwegian-forces/

12. 2020 Visit: Working the Integrated Distributed Insertion Force

1. https://www.2ndmardiv.marines.mil/News/Press-Releases/Article/2406055/ii-mef-conducts-regimental-air-assault-during-exercise-deep-water-20/

14. 2024 Visit: F-35s and Distributed Operations

1. https://www.2ndmaw.marines.mil/News/Article-View/Article/3670353/vmfa-542-becomes-first-f-35b-operational-squadron-on-east-coast-to-achieve-init/
2. https://defense.info/interview-of-the-week/major-brian-flubes-hansell-mawts-1-f-35-division-head/
3. https://nationalinterest.org/blog/buzz/madl-how-f-35s-talk-each-other-clear-game-changer-193119
4. https://sldinfo.com/2025/03/building-blocks-for-the-usmc-force-distribution-approach/
5. https://defense.info/multi-domain-dynamics/2025/04/the-heavy-lift-heli copter-and-its-role-in-supporting-diverse-usmc-operations/

15. 2025 Visit: Working Distributed Air Operations

1. https://www.dvidshub.net/video/886239/us-marines-use-tail-tail-method-conduct-logistical-operations

Made in the USA
Coppell, TX
07 November 2025